The Singapore Water Story

Singapore's journey during the past 45 years is an outstanding example that, in spite of multiple hardships, pragmatic policies, clear visions, long-term planning, forward-looking strategies and political will, as well as a relentless urge to improve, can result in strong foundations for sustainable development.

This book describes the journey of Singapore's development and the fundamental role that water has had in shaping it. What makes this case so unique is that the quest for self-sufficiency in terms of water availability in a fast-changing urban context has been crucial to the way development policies and agendas have been planned throughout the years. The authors analyze plans, policies, institutions, laws and regulations, water demand and water supply strategies, water quality and water conservation considerations, partnerships and importance of the media. They assess overall how all these issues have evolved in response to the dynamic needs of the city-state.

The study of Singapore shows how a dynamic society can address development without losing its focus on the environment. In the city-state, environmental concerns in general, and water concerns in particular, have played a major role in its transformation from a third world to a first world country. How and why this transformation took place is the main focus of this authoritative book.

Cecilia Tortajada is President, Third World Centre for Water Management, Mexico.

Yugal Joshi works with Northern Railways, Delhi, India, and formerly as a Research Fellow at the Lee Kuan Yew School of Public Policy, National University of Singapore.

Asit K. Biswas is Distinguished Visiting Professor, Lee Kuan Yew School of Public Policy, National University of Singapore and founder of the Third World Centre for Water Management, Mexico.

"*The Singapore Water Story* is a story of how Singapore successfully draws on clear vision, long term planning, constant innovation, a practical and effective approach and a national commitment to overcome the challenge of limited fresh water resources to develop a sustainable water strategy. In this remarkable book, the co-authors authoritatively and objectively analyse the various success factors in Singapore's sustainable water strategy and draws valuable lessons from Singapore's experience. It is an essential read for all water practitioners."

Yong Soon Tan, Permanent Secretary, Ministry of the Environment and Water Resources, Singapore, 2004–2010

"Water development horizons are beyond the life of most governments. The book's comprehensive story of Singapore illustrates the importance of consistent long term policies based on a holistic approach. Singapore has achieved sustainability through continuity of multiple supply source strategies and, very importantly, demand side management policies including investment to minimise non-revenue water."

Michael Rouse, Professor, University of Oxford

"*The Singapore Water Story* shows that Leadership and Commitment right from the highest levels in government down to the utility are critical for success. Empowering a capable workforce at the Public Utilities Board translated vision into outputs, 'safe drinking water, proper sanitation, and clean rivers' in Singapore. This book is a valuable knowledge product and recommended for wider knowledge sharing."

Bindu N. Lohani, Vice President, Knowledge Management and Sustainable Development, Asian Development Bank

The Singapore Water Story

Sustainable Development in
an Urban City-State

Cecilia Tortajada, Yugal Joshi and
Asit K. Biswas

Routledge
Taylor & Francis Group

LONDON AND NEW YORK

First published 2013
by Routledge
2 Park Square, Milton Park, Abingdon, Oxon, OX14 4RN

Simultaneously published in the USA and Canada
by Routledge
711 Third Avenue, New York, NY 10017

Routledge is an imprint of the Taylor & Francis Group, an informa business

British Library Cataloguing in Publication Data
A catalogue record for this book is available from the British Library

Library of Congress Cataloging-in-Publication Data
    The Singapore water story : sustainable development in an urban city state /
    edited by Cecilia Tortajada, Yugal Kishore Joshi, and Asit K. Biswas.
    pages cm
    Includes bibliographical references and index.
    1. Municipal water supply—Singapore. 2. Water-supply—Singapore.
    3. Water resources development—Singapore. I. Tortajada, Cecilia.
    II. Joshi, Yugal Kishore. III. Biswas, Asit K.
    TD313.S53S56 2013
    363.6'1095957—dc23
    2012039843

ISBN13: 978-0-415-65782-2 (hbk)
ISBN13: 978-0-415-65783-9 (pbk)
ISBN13: 978-0-203-07649-1 (ebk)

Typeset in Goudy
by Swales & Willis Ltd, Exeter, Devon

Printed and bound in Singapore by Markono Print Media Pte Ltd

Contents

Illustrations

Figures

Tables

Foreword

Life is full of improbable stories. There are also very improbable stories. And then, of course, there are the truly miraculous, nearly impossible stories. The Singapore Water Story clearly belongs to this category. This is why Cecilia, Yugal and Asit have done the world a big favour by documenting this remarkable story.

Why is this story miraculous? Just start with the basic facts. Singapore reluctantly assumed independence from Malaysia in 1965. Its reluctance was understandable. When a city is cut-off from its hinterland, it's like a heart cut off from its body. Yet Singapore survived and prospered as an economy because it connected its economic arteries to the rest of the world. But it could not get water from the world. It got it only from Malaysia on the basis of two agreements with clear expiry dates.

If anyone had suggested in 1965 that within fifty years Singapore could dream of achieving water self-sufficiency, he or she would have been laughed out of court. As the world's most densely populated country and as the world's only island city-state, water self-sufficiency would have been seen as an impossible goal. Yet this impossible goal is on its way to being realized (although Singapore values very much the continuing water agreement with Malaysia which expires in 2061).

The three authors are therefore right in suggesting in the Introduction that "Singapore has constructed one of the best public policy laboratories in the world in terms of water resources management". There are many secrets revealed in this book on how Singapore dramatically transformed its water management. But the real big secret is how Singapore integrated its public policies from many dimensions – from public housing to urban management, from industrial regulation to education. All this was done within a larger and complex political vision, which included a long-term view of Singapore's prospects. In addition, this vision had to be communicated to the people and bought by them. In short, it is a large, complex story. This volume brings out well the complexity.

The world turned a significant corner in 2008 when, for the first time in human history, more people lived in cities than in rural areas. Since then urbanization has accelerated even more, especially in Asia. Of the 25 most densely populated cities in the world, 17 are in Asia, and by 2025, 21 of the world's 37 megacities will be in Asia. Equally importantly, there are more than 450 cities globally with populations exceeding one million. Sadly, a number of these cities are not managing their environments well. The lessons from Singapore are clearly relevant to all these cities as many despair about providing clear and reliable water supply to all their inhabitants.

Singapore too was once a typical third world city. When I grew up in Singapore in the 1950s, we had no flush toilet at home. Today, virtually every home does. If Singapore could go from third world to first world in water management in one generation, it provides hope to millions, if not billions, of urban residents in Asia and beyond.

It is therefore clear that the Singapore Water Story needs to be studied and understood in depth by the people of Singapore and the world at large. The Lee Kuan Yew School of Public Policy and the Institute of Water Policy of our School are proud to have collaborated with the authors in this project. We would also like to offer our warmest congratulations to Cecilia, Yugal and Asit on completing this excellent story. We wish them great success for this project and for future projects that we will collaborate on. We also hope that this book will help to fulfill the mission of our School: to improve lives of millions, if not billions, by promoting better public policies, including those in the field of water management.

Kishore Mahbubani
Dean
Lee Kuan Yew School of Public Policy
National University of Singapore

Preface

In 2006, the United Nations Development Programme (UNDP) invited our Centre, the Third World Centre for Water Management in Mexico, to contribute to the 2006 Human Development Report (HDR) on "Beyond Scarcity: Power, Poverty and the Global Water Crisis". For this report, our Centre prepared and coordinated 20 case studies on different aspects of water management with experiences from all over the world.

One of the case studies focused on best practices of urban water and wastewater management at the global level for which one of the cities we thought about was Singapore. We thus contacted the then Chief Executive of the Public Utilities Board (PUB), Khoo Teng Chye, whom we did not know at that time. We explained to him the objective of our study and asked him if he could provide us with series of data which we would analyze and which would help us to decide if Singapore could be the focus of our case study.

Our further analyzes made us realize the outstanding performance of the PUB and, clearly, the fact that Singapore has one of the best urban water and wastewater management records in the world. Our study was included in the HDR, posted in full on the UNDP HDR website and later published in the *International Journal of Water Resources Development* in its June 2006 issue. Given the achievements of the PUB, we decided to nominate it for the Stockholm Water Industry Award. Not surprisingly, the Board was awarded this very prestigious prize.

During the course of our study, there were two issues that intrigued us: how Singapore had managed to transform its very poor urban water management system to become one of the best in the world within a short time span, and what had been the enabling conditions and the overall environment that had made this remarkable transformation possible. Our view was that if we could answer these questions objectively and satisfactorily, perhaps many other cities in both developed and developing countries would be able to learn lessons that had resulted from this process. We realized that they could use them with appropriate modifications depending on the specific social, economic, institutional and legal contexts.

The Singapore urban water study was the beginning of a very fruitful and enjoyable relationship between our Centre and the city-state. This collaboration led to the establishment of the Institute of Water Policy within the Lee Kuan Yew School of Public Policy, National University of Singapore.

Shortly after the Institute of Water Policy was established, our Centre proposed a collaborative research project, the objective of which would be to study how Singapore had managed to transform its urban water management practices so successfully when numerous cities all over the world have found this to be a truly difficult challenge.

We thus became very enthusiastic in studying Singapore's water policy, management, development and governance challenges, and how they have been overcome, twisted around and turned into development engines. During this process, we came across a wide array of information, insights and knowledge on how the city-state's urban water management has evolved over the last five decades and how water has played a key role in the overall development of the city-state. This book, which we have called *The Singapore Water Story*, is the end result of this research.

Throughout this study, we have been impressed by Singapore's culture of long-term and visionary planning, pragmatic decision-making, and the extent to which political will has backstopped its many strategic initiatives as part of its truly relentless pursuit of progress. Commendably, in spite of being endowed with few natural resources, the city-state has not seen its overall development thwarted. On the contrary, it has chosen and continuously tailored a path that has allowed it to grow and prosper. It has proven wrong all claims that such a small country would find it almost impossible to achieve so much and in numerous areas (including that of water resources) within a short period of time.

This book has only been possible thanks to the support and collaboration of a large number of institutions and people who contributed both directly and indirectly with their insights, experiences, comments and support. We are most grateful to all of them for their generosity and for sharing their knowledge, expertise and time despite their busy agendas and numerous professional commitments.

This study has been possible because of the strong support of the Lee Kuan Yew School of Public Policy (LKYSPP) at the National University of Singapore, and the Third World Centre for Water Management in Mexico. We are especially grateful for the continuous support of the Dean Kishore Mahbubani throughout this study.

We would also like to acknowledge the great support of former Minister of the Environment and Water Resources, Dr. Yaacob Ibrahim, as well as the PUB's former and present chief executives, Khoo Teng Chye and Chew Men Leong, respectively, for their immense help in facilitating access to data and information, organizing field visits, and offering invaluable insights on water management in the city-state as well as for opening doors to many personalities in Singapore who contributed to this process. We appreciate the valuable assistance offered to us by all of the PUB's staff during the course of this study, specially the Deputy Chief Executive, Chan Yoon Kum; Assistant Chief Executive, Policy, Chua Soon Guan; and Plant Manager at the Bedok NEWater Factory, Melanie Tan Li Ling. Their support deserves a special mention because it was critical for the study. We also want to extend our gratitude to the personnel at various PUB departments, mainly that of Policy & Planning; Industry Development; Technology; Water

Hub; 3P Network; Water Systems group; Catchment and Waterways; Water Supply Plants and Network; Water Reclamation Plants and Network; the Singapore International Water Week secretariat, etc.

At the beginning of the study, we were very fortunate to have two sessions lasting some three hours with former Prime Minister and Minister Mentor Lee Kuan Yew. We received from him a first-hand account of the countless difficulties Singapore encountered before and after its independence and the reasons why the city-state took so many fundamental early water-related decisions. His visionary approach, forward-looking views and in-depth comments were critical for us to understand the extraordinary effort behind Singapore's transformation in its quest for sustainable development and the roles water played in the transformation process of the city-state. We are highly indebted to former Prime Minister Lee Kuan Yew for his generosity.

We were fortunate to interview other several outstanding personalities of Singapore as well who have tirelessly and energetically pursued the development of the city-state. These include PUB Chairman Tan Gee Paw, whose remarkable foresight and continuous support guided us in this learning exercise. Our conversations with former PUB Chairman Lee Ek Tieng, responsible for the cleaning of the Singapore River, provided us with further insights. Equally valuable were the reflections of Director, RSP Architects Planners & Engineers (Pte) Ltd., and former Chief Executive Officer of the Housing & Development Board (HDB) Liu Thai Ker, and Senior Vice President of Singbridge International Pte Ltd., and former Group Director of Singapore's Urban Redevelopment Authority International Wong Kai Yeng.

The continuous encouragement and advice of former Permanent Secretary for Environment and Water Resources, Tan Yong Soon, were especially important as he shared with us his knowledge and broad views.

Our appreciation also goes to Dr K.E. Seetharam, former Director of the Institute of Water Policy, LKYSPP; Industry Development Director Michael Toh, PUB; and Singapore International Water Week (SIWW) Managing Director Maurice Neo for their support; as well as former Water Hub Manager Tan Ban Thong for his continued insights; URA's former Chief Executive Cheong Koon Hean; Prof. Lee Poh Onn, Institute of Southeast Asian Studies; Prof Cherian George, Wee Kim Wee School of Communication and Information, Nanyang Technological University; Dr Leong Ching, LKYSPP; Kimberly Pobre, LKYSPP; and Prof. Goh P.S. Daniel, Department of Sociology, National University of Singapore. We are also indebted to numerous experts at the PUB, the National Environment Agency, the Ministry of Foreign Affairs and the Urban Redevelopment Authority for commenting on the draft manuscript.

Personnel at the different libraries and resource centres we consulted were extremely helpful. In Singapore, these centres include the National University of Singapore (C. J. Koh Law Library, Central Library and Science Library); the Information Resource Centre, Singapore Press Holdings; Resource Centres at the PUB's Headquarters and the NEWater Visitor Centre; the National Archives; the National Library; and the Resource Centre at the Urban Redevelopment

Authority Centre. In the United Kingdom, the National Archives and the British Library; and, in the United States, the Library of the Congress and the New York Public Library.

In our search for documentation, we stumbled upon the vast archives held at NUS Libraries, which in our opinion have one of the best collections in the world. These Libraries are truly a gem for all researchers to explore irrespective of the topic on which they are working.

We would also like to thank John and Teresa Chamberlain for sharing with us their photographic archive and allowing us to use some of the images of Singapore back in the 1970s when they lived in the city-state. We would also like to express our appreciation to Andrea Lucia Biswas-Tortajada for her numerous suggestions for improvement of the manuscript and to Thania Gomez for her secretarial and other support. Last, but not least, we would like to thank Routledge and the team working with Tim Hardwick for all their patience and support. Without all this support the study and the book simply would not have been possible.

Cecilia Tortajada
Yugal Joshi
Asit K. Biswas
Singapore, September 2012

Abbreviations

ABC Waters	Active, Beautiful and Clean Waters
ADB	Asian Development Bank
APU	Anti-Pollution Unit
BBC	British Broadcasting Corporation
BCD	Building Control Division
BH	*Berita Harian*
BM	*Berita Minggu*
BO	Biochemical Oxygen
BOD	Biochemical Oxygen Demand
BP	*Birmingham Post*
BT	*The Business Times*
BW	*Business World*
CBPU	Central Building Plan Unit
CCC	Citizens' Consultative Committee
CKS	Centre for Khmer Studies
CPOOC	Code of Practice on Pollution Control
CWO	Corrective Work Order
DBOO	Design-Build-Own-Operate
DGP	Development Guide Plans
DNPU	Direct Non-Potable Use
DO	Dissolved Oxygen
DPA	*Deutsche Presse-Agentur*
DPVG	Dengue Prevention Volunteer Group
ECAFE	Economic Commission for Asia and the Far East
EDB	Economic Development Board
EEA	Environmental Education Advisor
EHD	Environmental Health Department
EIC	(British) East India Company
ENV	Ministry of the Environment
EPCA	Environmental Pollution Control Act
EPHA	Environmental Public Health Act
EPMA	Environmental Protection and Management Act
FM	Foreign Minister

GDP	Gross Domestic Product
GST	Government Service Tax
HDB	Housing & Development Board
IHT	*International Herald Tribune*
IIAS	International Institute for Asian Studies
IMCSD	Inter-Ministerial Committee on Sustainable Development
IND	Indonesia
IPU	Indirect Potable Use
ISEAS	Institute of Southeast Asian Studies
JTC	Jurong Town Corporation
KL	Kuala Lumpur
LCFC	Low Capacity Flushing Cistern
LKY	Lee Kuan Yew
LRT	Light Rail Transit
MCA	Malaysian Chinese Association
MEWR	Ministry of the Environment and Water Resources
MFAS	Ministry of Foreign Affairs of Singapore
MIC	Malaysian Indian Congress
MICA	Ministry of Information, Communications and the Arts
MM	Minister Mentor
MM	*Mingguan Malaysia*
MND	Ministry of National Development
MP	Member of Parliament
MRT	Mass Rapid Transit
MTI	Ministry of Trade and Industry
MWELS	Mandatory Water Efficiency Labelling Scheme
MY	Malaysia
NCMP	Non-Constituency Member of Parliament
NEA	National Environment Agency
NEAC	National Economic Action Council
NMP	Nominated Member of Parliament
NSP	*Nanyang Siang Pau*
NST	*New Straits Times*
NSTP	*New Straits Times Press*
NUS	National University of Singapore
NW	*Newsweek*
ODN	*Oriental Daily News*
OECD	Organisation for Economic Co-operation and Development
PA	People's Association
PAP	People's Action Party
PCD	Pollution Control Department
PM	Prime Minister
POA	Points of Agreement
PPD	Primary Production Department
PSA	Port of Singapore Authority

PU	Public Utilities
PUB	Public Utilities Board
PWD	Public Works Department
RTS	Radio and Television Singapore
SBS	Singapore Broadcasting Corporation
SCD	*Sin Chew Daily*
SDA	Sewerage and Drainage Act
SG	Singapore
SIT	Singapore Improvement Trust
SIWW	Singapore International Water Week
SLOA	Singapore Lighter Owners Association
SM	Senior Minister
SPH	Singapore Press Holdings
ST	*The Straits Times*
STB	Singapore Tourism Board
SUI	Singapore Utilities International
TE	*The Economist*
TI	*The Independent*
TKH	*The Korean Herald*
TNW	*The Nikkei Weekly*
TSS	Total Suspended Solids
UM	*Utusan Malaysia*
UMNO	United Malays National Organisation
UMP	Utusan Melayu Press
UN	United Nations
UNDP	United Nations Development Programme
UNESCO	United Nations Educational, Scientific and Cultural Organisation
URA	Urban Redevelopment Authority
WEB	Water Efficient Building
WEH	Water Efficient Homes
WELS	Water Efficiency Labelling Scheme
WEMP	Water Efficiency Management Plan
WHO	World Health Organization
WN	*Waste News*
WPCDA	Water Pollution Control and Drainage Act
WVG	Water Volunteer Group
WWS	Waterways Watch Society

Introduction

The book presents Singapore's journey to plan, manage, develop and govern its water resources from the time of its independence in 1965. At that time, social, economical, political and environmental constraints were such that those looking from outside the city-state would have most likely predicted a bleak future. Reality, however, turned out to be very different due to the determination and the visionary initiatives of Singapore's leaders not only to survive but also, and quite commendably, to strive. This constant drive to excel and reach ever higher standards of living has propelled the city-state to move radically away from all the dire development predictions that were once made about its future. Instead, today Singapore stands as an example of widespread resilience, innovation, resourcefulness and determination.

At the time of independence, separation from Malaysia brought a myriad of complex problems. As a newly independent nation, Singapore had to suddenly face a future with no hinterland, no natural resources and almost total dependence on Malaysia for water supplies. When the two countries parted ways, Singapore worried about whether previously signed agreements to regulate the supply of water from Malaysia to Singapore would continue to be honoured and wisely sought ways to entrench them.

With an area of 580 km^2 at the time of independence and not enough land to collect but a small amount of the 2,340 mm of average rainfall it gets every year, the city-state had to expediently devise strategies to develop and manage its water resources. These plans reflect innovative, creative and forward-looking initiatives to ensure that it would have the water resources it needed to satisfy the increasingly diverse needs of a growing population; fuel economic development to generate employment; and ensure a healthy and pleasant natural environment.

With a proactive development approach and in the search for the best public policy, management, technical and technological solutions, Singapore's policy-makers have made impressive achievements towards reducing its reliance on outside sources and strengthening its own internal capacities. For example, at the time of independence total water consumption stood at 70 million gallons per day (Mgal/day), there were only three reservoirs and the catchment area covered only 11 per cent of the city-state. By 2011, the number of reservoirs had increased to 17. The most impressive of them, Marina Reservoir, is located in the city centre, at the pinnacle of urbanization and at the heart of the business and commercial hubs. Its catchment area covers 10,000 ha or one-sixth the size of the city-state, and, together with some other reservoirs, has increased Singapore's water catchment from half to two-thirds of the country's land area.

Throughout the years, consistent and considerable investment has been poured into developing not only conventional sources of water but also unconventional ones. As such, desalination, NEWater (treated wastewater or used water as it is known locally) and industrial water capacities have all gone from being non-existent in 1965 to representing 30, 117 and 27.5 Mgal/day, respectively, by 2011. In fact, it has been the development of intuitive, known and feasible alternatives and the visionary exploration and innovation of options that were deemed and often thought of as unfeasible and well beyond the obvious, which have opened up an array of flexible and feasible water supply alternatives for the city-state.

Given its geographical conditions, it is not surprising to see how the quest for water self-sufficiency has put considerable strain on a land-constrained Singapore. The pursuit of this aim has also had a significant impact on the island's urban development, which has translated into a comprehensive, coordinated and forward-looking way in which urban development has been conceived, planned and carried out. In Singapore, development came to be equated with urbanization and this process had to factor in water if it was to succeed. The resulting experiences have made this a valuable example of how a city-state is able to formulate long-term plans, implement them in a timely and cost-effective manner, and keep on moving towards sustainable development as part of its overall growth strategies and development path.

Although Singapore cannot be compared to many cities due to its unique position as an island nation and having only one level of government, it has certainly faced many of the same challenges and problems of any entity experiencing rapid industrialization and urbanization. If anything, addressing and resolving most difficulties has proven even more critical for the city-state due to the lack of hinterland to rely on for natural resources and the state of dependence to draw part of its water resources from another country. This has meant that the small island has always absorbed all by itself the adverse impacts of growth on the environment. It has also meant that all development-related decisions have been traditionally dealt with as being of utmost importance by all ministries. Fully backed, led and promoted by the highest political level, strategies aiming at propelling the city-state ahead have been formulated with a common goal in mind, fully aware of the fact that decisions on one social, economic, financial or environmental issue most likely will have a bearing in one or more of the others.

This level of pragmatism and non-political policies and strategies have led to the creation and strengthening of the sort of physical, institutional and human capital infrastructure that has been seized by foreign investment and greatly contributed to the city-state's impressive economic development record. This material prosperity has allowed for the resources that have in turn become the backbone of Singapore's nation-building programme, where the Public Utilities Board (PUB) has played a very important role. The Board has made sure that services are efficiently delivered even before the needs emerge to cover the increasing demands of domestic, commercial and industrial sectors as well as for nature and recreational uses, both at present and looking towards the future.

One could say that innovative water resources planning and management have resulted out of necessity given their naturally limited availability and thus the constraints the city-state has faced. Nevertheless, the policies, programmes and plans that have emerged from this imperative have come from a clear national vision and long-term coordinated and continuous planning exercises where the city-state's overarching goals have prevailed over those of any of its individual sectors irrespective of their relevance. The overall vision has always received significant and constant support from the highest ranks of the political life. It is thus appropriate to stress that one of the most important lessons Singapore can teach developed and developing countries alike is the exemplary political will of its leadership. Even when decision-making has been heavily centralized, measures have remained matter-of-fact, not ideological and in line with the constant quest for the best long-term and lasting benefits for the city-state and its people. Both the role of the leadership and the importance of inter- and intra-institutional coordination are definitive aspects that will recurrently come across in each one of this book's chapters as the consistent strengths of the city-state.

Throughout the years, it has been the team at the front of the government who has envisioned and encouraged a sustained process of social and economic development through which the quality of life of the population has been significantly improved, the environment has been protected and the city-state has placed itself on the right track towards sustainability. In 1960s, the visionary Prime Minister, Lee Kuan Yew, realized that the only way to achieve economic, social and environmental gains for the people of Singapore was through the formulation of integrated, holistic long-term policies that promoted coordination among the different ministries, agencies and sectors. It was clear to him that these efforts were worth pursuing in spite of the complexity to implement them. To the benefit of Singapore and its people, this approach to problem-solving and decision-making prevails as a rule until the present day.

In the chapters included in this book, the reader will be able to identify how holistic policies and management practices for water resources have been developed and tailored according to the city-state's specific endowments and changing needs, always with a long-term perspective and a lasting positive impact in mind. For the sake of the common good, and despite their intricacy, systems for vertical and horizontal coordination, cooperation and communication between different ministries, agencies and actors have been put in place and sustained throughout the years. It is important to mention that all dollar figures are quoted in Singapore dollars unless otherwise specified.

Chapter 1 focuses on water supply strategies by setting the scene and their foundation. It addresses the way water resources have been developed since 1965, the decisions and the rationale as to why they were made, the historical situations that prompted them and what strategies have been developed until today.

Chapter 2 then goes on to discuss how water has been incorporated as a key factor in Singapore's urban development process in the context of rapid urbanization and industrialization, an experience that stands as a valuable case study in the area of sustainable urban development.

The stringent and evolving legal and regulatory frameworks that have been developed in the city-state to control increasing water, but also air and land pollution, are analyzed in Chapter 3. This topic is particularly relevant because Singapore has been able to attract significant amounts of foreign investment for industrial development in spite of, and perhaps also because of, its strict environmental regulations and enforcement schemes. This has contradicted claims made by every other nation that a sturdy environmental protection framework will drive international investors away and will thus have negative impacts on the future of industrial development.

The management of water demands as part of national security considerations is dealt with in Chapter 4 in a framework of increasing domestic and industrial uses as well as for nature and recreational activities. This section discusses how pricing and mandatory water conservation strategies have been implemented. It also argues that further priority should be awarded to economic mechanisms to reshape consumption habits and patterns and encourage more rational consumption of water.

Closely related to influencing behaviour around using water, Chapter 5 discusses the different education and information strategies that have been implemented for conservation purposes. It also presents an account that the reader may find useful to understand the unique water relationship that exists between Singapore's civil society and the government and how this interaction has reflected onto the many campaigns seeking to engage and reach out to the general public over recent decades.

Chapter 6 presents the strategies implemented for the cleaning-up of the several river systems in Singapore, many of which were significantly polluted when the city-state became independent. These ten-year-long activities were part of the redevelopment of the city-area and included massive infrastructural and public service development as well as the relocation of large parts of the population and reorientation of their economic activities. This exercise has proved key for the urbanization and the rehabilitation of run-down areas by injecting such neglected sites with the natural and man-made landscape required to draw in a revitalizing commercial community and become a vibrant place for recreational activities.

Chapter 7 focuses on the views of the media on the water relationship between Singapore and Malaysia. A relevant player in shaping the water relationship, the media has acted both as a reporter and as a vehicle of communication, both officially and unofficially, to its own public and also to the interested parties in the other country.

Finally, Chapter 8 looks ahead to Singapore's water and sustainable development prospects. It discusses its present and possible future challenges from a water security viewpoint in an environment of uncertainties under scenarios where, at the national, regional and global levels, the only certainty is change.

With this book, we hope readers both within and outside Singapore will be able to learn from the manifold experiences the city-state has to offer. In fact, by looking at its journey towards sustainable development through water lenses, it can be concluded that the holistic management of water resources has paved the way for the progress so far attained towards its ambitious development goals. It is clear that water resources have been crucial to Singapore's sustainable development and the improved quality of life of its population, a key role that will only be heightened in the future. This book stresses how this has been the case.

Like in all things, there is always room for improvement. Nevertheless, so far, Singapore has constructed one of the best public policy laboratories in the world in terms of water resources management. Both developed and developing countries can certainly draw from the lessons learned and emulate those policies and practices that may prove relevant. The chapters ahead emphasize some of the most meaningful ones.

1 Setting the foundations

Introduction

Singapore was founded in 1819 by Sir Stamford Raffles of the British East India Company (EIC). Due to its strategic location along the Straits of Malacca, Raffles decided to make of the island a free port where entrepôt trade could flourish. By the late eighteenth century, Singapore had become one of the world's most important and busiest ports. With the development of steamships (1840s) and the opening of the Suez Canal (1869) the volume of trade increased and the economy flourished (Turnbull, 2009).

The island's main trade artery was the Singapore River flowing to the south, and most of the city developed outwards from its banks in what came to be known as the central area. This area was administered by the City Council while the Rural Board administered the rural areas.

Decades of rapid development and lack of long-term planning resulted in overcrowding in the central area. Upon the establishment of self-government in 1959, population density in this area reached over 2,500 inhabitants/ha as people occupied vacant and marginal lands and lived in flammable huts without water, sanitation or any elementary public health service whatsoever (PUB, 1985b; Tan, 1972). At this time, the government faced a multitude of challenges. Nation-building and the economy were of utmost importance, but there were also pressing issues related to the reorganization of governmental administrative organs, including the absorption of the City Council departments into government ministries, labour, education, trade unions and social security, housing, rural development, health, women, etc. (Toh, 1959). Moreover, the communist threat in South-east Asia came to significantly add to the difficulties Singapore was already facing.

Central to nation-building was the need to improve the population's quality of life and provide them with a sense of ownership. The ruling party, the People's Action Party (PAP),[1] developed a five-year plan to address the disparities in the quality of life of the population living in urban and rural areas. In the central area, problems related to economic activities, lack of housing and overcrowding had to be addressed. On the other hand, rural communities needed to be provided with basic amenities and services including electricity, water pipes and drainage schemes, modern kampong (village) transportation networks, health facilities for women and children, hospitals, and adequately equipped community centres.

On 1 February 1960, the Housing & Development Board (HDB) became the country's first statutory body established to tackle overcrowding and the serious housing shortage. This was followed by the creation of the People's Association (PA) on 1 July 1960 to deal with communist and communal threats by controlling and coordinating the 28 community centres it was to oversee. One and a half years later, the Economic Development Board (EDB) was established mainly to attract foreign investment and solve problems with growing unemployment (Quah, 2010).

The City Council and the creation of the Public Utilities Board

The Municipal Commission was established in 1887 and was responsible for supplying piped water to the public. It subsequently took on the role of supplying gas and electricity too, in 1902 and 1906 respectively (PUB, 1985a). When the British granted Singapore city status in 1951, the Municipal Commission changed its name to the City Council. At that time, the government structure consisted of a local government (formed by the City Council and the Rural Board) and a central one.

In 1957, the tasks of the City Council were transferred to the Ministry of Local Government. During the 1959 elections, the PAP recognized the duplication of efforts within the City Council and other ministries and committed, if elected, to merge several of this body's departments into the government service in order to increase efficiency, improve administration and achieve financial stability.

This became a main element of the PAP platform during the elections as bureaucracy in Singapore was a serious problem. To give an example of bureaucratic red tape, Inche Baharuddin Bin Mohamed Ariff, Member of Parliament for the Anson district, made the following statement during a Parliament debate on the Suspension and Transfer of Functions Bill for the City Council:

> The best example I can cite is the question of building houses. The plans for these houses are given to the City Council, then they are given to the Department of Lands and then to the S.I.T. [Singapore Improvement Trust]. If they are concerned with rural areas, the plans are sent to the Rural Board.
> (Ariff, 1959: column 60)

In terms of water, the then Minister of National Development, Ong Eng Guan, made the following comment:

> For a village in the rural area to get a standpipe, the headman must first write to the Rural Development Commission, which is under the direct jurisdiction of the Chief Minister, the present Member for Cairnhill. The Chief Minister will write to the Member of the Local Government, perhaps, who in turn writes to the Rural Board, and which might in turn write to the

District Council, which in turn writes to the City Council, perhaps, through the auspices of the Ministry of Local Government. Hundreds of memos would have been sent around asking for a standpipe to be erected in the village. The people in the kampong cannot wait for the Ministers and their Permanent Secretaries and the Heads of Departments wasting time on the memoranda for their needy water supply.

(Ong, 1959a: column 91)

Ong Eng Guan also provided an example of the bribery existing in the City Council:

Before we went into the City Council in December 1957, it was the general knowledge of hawkers[2] that, in order to get licences, the applicant must get the requisite number of bottles of 'TST' brandy, the requisite number of katties of oranges and tins of biscuits, and send them to the person concerned in order to get a licence.

(Ong, 1959a: column 92)

Therefore, in an effort to improve the government's efficiency, the City Council was suspended in 1959 and its functions gradually transferred to the Ministries of National Development and Local Government. To keep the ministries focused on their main strategic tasks and to reduce their workload, statutory boards were established to perform more operational functions.

The Public Utilities Board (PUB) was established in 1963 under the Prime Minister's Office and was made responsible for supplying water, electricity and gas (PUB, 1963). In turn, the PUB's Water Department was responsible for the collection, storage and treatment of raw water to drinking water; the operation and maintenance of the water supply pipe network; water delivery to the consumer; and water supply through standpipes and trucks. In November 1964, the PUB was transferred to the Ministry of Law (PUB, 1964), and taken back to the Prime Minister's Office in 1971 (PUB, 1971). In the 1980s, the PUB was part of the Ministry of Trade and Industry (MTI), before becoming a statutory board under the Ministry of the Environment (ENV) in April 2001, taking over the latter's Sewerage and Drainage Departments in a reorganization that reflects the completely integrated approach by which water is managed in Singapore. The ENV was then renamed as the Ministry of the Environment and Water Resources (MEWR) in 2004 (Tan et al., 2009).

From the time it was established, the PUB started developing the island's water supply system taking into consideration the increase in water demand resulting from economic development, a 30-fold rise in Gross Domestic Product (GDP), an over 20 per cent land augmentation due to land reclamation, and a tripling population. In order to respond to the needs as well as anticipate the challenges of a growing country, Singapore's water resources planning, management, development and governance strategies have been developed in what has become one of the best systems all over the world. Table 1.1 shows some key statistics describing

Table 1.1 Key statistics on Singapore, 1965 and 2011

	1965	2011	Change
Land area (km²)	580 km²	714 km²	+134 km²
Population	1,887,000	5,184,000	3,297,000
GDP per capita[a]	$1,580	$63,050	$61,470
Water consumption per capita l/person/day	75 l/person/day	153 l/person/day	+78
Total water consumption[b]	70 Mgal/day	380 Mgal/day	+310 Mgal/day
No. of reservoirs	3	17	+14
Land area as water catchment	11%	67%	+56%
Desalination capacity	0	30 Mgal/day	+30 Mgal/day
NEWater capacity	0	117 Mgal/day	+117 Mgal/day
Industrial water capacity[c]	0	15 Mgal/day (2010)	+15 Mgal/day
Water availability	24 hours/day	24 hours/day	–
Service coverage	~80%	100%	
Unaccounted-for-water	8.9%	5.0%	–3.9%

Sources: MEWR, 'Key Environment Statistics – Water Resource Management', Singapore. Available at: http://app.mewr.gov.sg/web/Contents/contents.aspx?ContId=682 (accessed on 22 August 2012); Department of Statistics 'Key Annual Indicators', Singapore. Available at: http://www.singstat.gov.sg/stats/keyind.html (accessed on 22 August 2012); PUB Annual Reports.

Notes:
a At 2011 market prices.
b Water demand shown here is for 2011, while water supply capacity includes the fifth and latest 50 Mgal/day NEWater plant at Changi.
c Assume industrial water capacity = sales of industrial water. Information extracted from http://app.mewr.gov.sg/web/Contents/Contents.aspx?ContId=682 (accessed on 9 March 2012).

how the city-state has been transformed over the last four decades, mostly from the water resources perspective.

In the following the early state of Singapore's water resources development is discussed, as well as the water supply plans that were conceived in 1950, 1962 and 1972 respectively. These three latter plans have set the foundations for reliable and clean water supply that has played a very important role to ensure sustained economic and social growth in the city-state.

Early stage of water resources development

When Raffles landed in Singapore in 1819, water from inland streams and self-dug wells was enough to maintain the island's 150 or so inhabitants. As trade-related activities increased, it became urgent to supply water to the ships calling at its port and so a small reservoir was constructed at Fort Canning in 1822. As the port developed, and as early as 1890, overcrowding and pollution had contaminated most of the wells in the city, making them unsafe for drinking. The wells were therefore closed down in the 1890s, but were reopened in 1902 once again due to increasing water demand (PUB, 2002b).

Measures taken to increase water supply included the construction of an earthen dam in the Kallang Valley in 1867. As the dam proved unsatisfactory, two improvements were made to the reservoir between 1890 and 1904, moving the dam to its present location and raising it by 1.5 m (PUB, 1985b). In 1922, the modified project was named MacRitchie Reservoir after Municipal Engineer James MacRitchie, who oversaw the expansion works and who was behind the development of improving schemes for the then town of Singapore.

Water was still in short supply, with dry spells occurring in 1877 and 1895 (Ooi, n.d.). To alleviate water shortages, the Singapore Municipality built the Kallang River Reservoir in 1910 and renamed it as Peirce Reservoir in 1922, after the municipal engineer in charge of its construction, Robert Peirce. The third impounding reservoir, Seletar, was built in 1920 within the central catchment and expanded in 1940 (PUB, 1985b). It was named after a Malay word that refers to coastal dwellers (Orang Seletar). The three reservoirs in the protected central catchment area had a total supply capacity of 17.5 Mgal/day and a total storage capacity of about 2,100 Mgal. They had a series of pumps to transfer surplus water from one reservoir to another to maximize yields. For instance, Seletar Reservoir had a storage capacity of 150 Mgal while Peirce had a much larger storage capacity of over 900 Mgal for which pumps were installed to transfer surplus water from Seletar to Peirce Reservoir.

By 1920, the population had increased to more than 400,000 people and the British began looking towards the mainland as a possible source of water. As a result, an agreement was signed in 1927 with the Sultan of Johor in Malaysia for the construction of the Gunong Pulai and Pontian reservoirs and export water to Singapore (see Agreement as to Certain Water Rights in Johore[3] between the Sultan of Johore and the Municipal Commissioners of the Town of Singapore signed in Johore on 5 December 1927). The Gunong Pulai and Pontian reservoirs became operational by 1932, supplying some 18 Mgal/day of water (PUB, 2002b).

Since the Second World War, MacRitchie, Peirce and Seletar reservoirs supplied Singapore's water together with Malaysian water imports from the Gunong Pulai and Pontian reservoirs. There also used to be a small saltwater system for public use, with a 0.5 Mgal/day capacity. However, the pumps were inefficient and the City Council decided to close it once there was enough freshwater available.

Water supply plans

The 1950 Water Resources Study

In 1950, the British commissioned a group of consulting engineers formed by Bruce White, Wolfe Barry and Partners to investigate the availability of additional water resources in the island and how these could be utilized to meet the increasing demand, as well as to develop an emergency water supply should water from Johor be cut off. The consultants noted that 'the possibility of circumstances arising which would cause an interruption in the supply from the mainland is a matter of concern to those responsible for the administration of the island' (White, 1950: 2). This study seems to have been commissioned drawing from the lessons learnt during Singapore's fall in the Second World War, when water to the city was cut off and the island became very vulnerable.

Water demand projections

At the time of the 1950 study, there were two main formal modes of water supply: direct piped water supply to households and piped water supply to standpipes for the community. The former was more common in the municipal or city areas where about 760,000 people lived; the latter was used to supply water to the approximately 240,000 inhabitants residing in villages outside the city area as well as some communities within the city limits. In addition to water supplied from standpipes, rural communities also made use of self-dug wells.

The consultants assessed water consumption in Singapore since 1819 and projected that its population would double within 20 years from one million in 1950 to two million in 1970. They also projected that per capita water consumption would increase from 140 l/capita/day in 1950 to about 225 l/capita/day, noting that an average per capita consumption of 140 l/capita/day was low for a tropical country. This figure was calculated dividing the island's total consumption by total population, therefore making no differentiation between domestic and non-domestic consumption.

Water supply options and recommendations

The consultants looked into and recommended three options to increase Singapore's water supplies: drawing water from rivers by either damming them or transferring water to a larger central reservoir; tapping on available groundwater; and constructing wells and harvesting rainwater from roofs.

A number of rivers in the west that could be turned into reservoirs were identified, namely the Sungei Sembawang, Kranji, Jurong, Pandan, Namly and Alexandra. Nevertheless, it was concluded that their river beds were not suitable for the construction of dams to create reservoirs. Instead, it was suggested to construct a central storage reservoir with a storage capacity of 600 Mgal, west of the existing protected central catchment, to where water from the rivers would be pumped. The total dry weather flow from the river sources was estimated at 19 Mgal/day.

The group also identified the Bedok area in the east as a potential site for tapping on an estimated potential yield of 10 Mgal/day from groundwater resources. This was perceived to be a conservative estimate as the consultants had assumed that only 10 per cent of the estimated 80–100 Mgal/day of water that entered the ground could be pumped for water supply. They did not quantify the amount of water supply that could be drawn from these sources but noted the shallow nature of existing wells. They also mentioned that it could be possible to make available a large supply of water if funds were allocated for the construction of wells with linings, parapets and covers. The alternative of collecting rainwater for domestic purposes from buildings was also suggested.

The government's general thinking regarding the development of water resources was that it would be more economical to obtain groundwater from a single source than to produce it from the several scattered rivers. Nonetheless, the consultants recommended that the water-bearing capacity in the east should be investigated, after which a decision could be taken on the need to draw on river supplies. Figure 1.1 shows the locations of the proposed water supply sources.

Security considerations

It was commonly perceived that Johor sources in Malaysia represented an almost unlimited supply of water and thus could meet the needs of a growing community. However, the consultants quoted Lieutenant-General Percival's official account on the fall of Singapore to Japan during the Second World War, noting that the problem with water supply was not lack of availability but the fact that heavy bombing had damaged water pipelines. These pipes could not be repaired quickly enough as all civil labour had disappeared, as a result of which more than half of the water had been lost. There were also considerations on whether additional water supplies (i.e. groundwater and augmented surface water sources) would be used to meet demand or be kept as a reserve for times of emergency.

Developments between 1950 and 1962

From 1950 to 1962, stemming from the 1950 Water Resources Study recommendations by White & Partners, most of the plans to increase local water supply sources

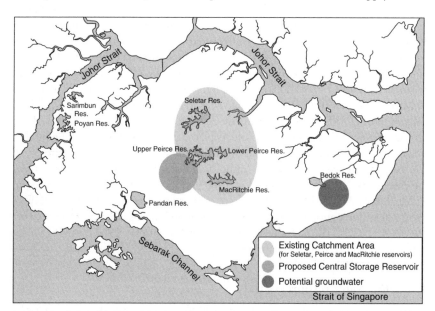

Figure 1.1 Map showing potential water supply sources in Singapore (White, 1950).

focused on exploring groundwater. The groundwater system at Bedok was commissioned in 1959 but the yield fell short of expectations with a maximum output in the region of 0.75 Mgal/day, instead of the estimated 10 Mgal/day. The then Chief Water Engineer, P.C. Lim, pointed out that the cost of treatment was much higher than for other works in Singapore and Johor for which Bedok might eventually 'prove to be a white elephant' (City Council of Singapore, 1959: 1). At the same time, the British were also studying plans to expand water supply capacity at the Malaysian Tebrau and Johor rivers in order to meet water demand in Singapore.

It is important to mention that, in the 1920s, the Johor River and Gunong Pulai and Pontian reservoirs were assessed as potential water supply sources for Singapore. However, at that time, the Johor River scheme was rejected in favour of the Gunong Pulai and Pontian schemes (Binnie & Partners, 1981). In the 1940s, the Johor River scheme was seriously considered and land even acquired, but the project had to be deferred because of the Second World War (Binnie & Partners, 1981; White, 1952). The Tebrau water supply scheme, which involved the construction of a raw water intake station and waterworks at Tebrau in Johor to abstract and treat raw water from Tebrau River, was later identified.

The Johor River scheme had an estimated yield of at least 100 Mgal/day and could potentially provide more water than that in Tebrau, the yield of which was assessed at 60 Mgal/day. However, the City Water Engineer recommended developing the Tebrau scheme first, primarily because the emergency conditions due to the communist insurgency in the Malayan jungles in Johor made it dangerous to construct the waterworks and related pipelines within the forested areas. In addition, the quantity of water available at Tebrau was sufficient to meet demand for at least 20 years. Furthermore, the cost of the Tebrau scheme was also calculated to be less expensive since it did not require a full treatment facility and the closer distance to Singapore shortened the required pipeline length. There were also proposals to treat raw water abstracted from Tebrau in Singapore instead of doing so in Johor to centralize treatment and reduce costs. Finally, Tebrau's waterworks began in the early 1950s and were completed in 1953 with an initial capacity of 10 Mgal/day which was expanded to 20 Mgal/day in 1954.

Later on, in 1958, the government decided to undertake the Johor River Project, thereby abandoning the projects planned under the Tebrau scheme that would become redundant (City Council of Singapore, 1959). It was presumed that the Johor River could supply 'unlimited' water but as this proved not to be the case, work on Tebrau resumed in 1960 and its capacity further expanded to 45 Mgal/day in 1962. Moreover, in spite of rises in water exports from Malaysia, there was still not enough water to meet Singapore's needs whenever the weather was dry. In fact, all through the September 1961 to January 1962 drought, Singapore was subject to water rationing.

The 1962 water resources study

In 1961, the City Council signed a Water Agreement with the State of Johor in Malaysia to give Singapore the 'full and exclusive right and liberty to take,

impound and use all the water' within the Gunong Pulai and Pontian catchments and Tebrau and Scudai rivers until 2011 (see Tebrau and Scudai Rivers
Agreement between the Government of the State of Johore and the City Council of the State of Singapore signed on 1 September 1961). The following year
another agreement was signed to supply up to 250 million gallons of water per day
to Singapore from the Johor River until the year 2061 (see Johore River Water
Agreement between the Johore State Government and City Council of Singapore signed in Johore on 29 September 1962).

Subsequently, in 1962, the Government of Singapore commissioned Binnie &
Partners to conduct a study that involved four areas: the economic and technical
appraisal of the existing and planned water supply works in Singapore and Johor;
determine the detailed design of the Johor River scheme to obtain a loan from
the World Bank; look into the necessity of developing supply works at Scudai as
an interim (to immediately meet demand while the Johor River Project was constructed) or a permanent scheme; and examine the proposal to create a freshwater
lake in the Johor Straits.

Water demand projections, options and recommendations

In their report, Binnie & Partners noted that the rate of increase in total water
consumption in Singapore had been fairly uniform since 1950. They thus extrap-
olated the rate of growth of total water demand to estimate demand at 142 Mgal/
day in 1982. This figure was compared with the estimated water demand based
on approximate growth in population (3,186,000 in 1982) and estimated increase
in per capita water consumption (46.35 gallons/capita/day or 210 l/capita/day).
Again, projections did not make any differentiation between domestic and non-
domestic water demand. Instead, figures assumed that future requirements from
major water-using industries would be met from resources developed by the then
Jurong Industrial Development Board and that other industrial demands on the
City Water Department could be covered by a modest increase in per capita consumption for the city as a whole.

At that time, serious thought was put into the expansion of the Seletar Reservoir instead of expanding capacity in Johor. There were several benefits of increasing Seletar Reservoir's capacity: the yield from Tebrau River could be augmented
by pumping raw water from the Tebrau to the Seletar Reservoir and raw water
from Johor River could also be sent to Seletar to be treated in Singapore instead of
building a new treatment works at Johor River. This would centralize treatment
capacity and would simplify administration and minimize costs. Finally, Seletar
Reservoir would provide additional storage for flood alleviation.

The consultants concluded that the increased exploitation of the Seletar catchment should be taken into consideration, but that this could not be
regarded as an alternative to the Scudai or Johor River schemes. They also suggested that the initial stage (30 Mgal/day) of the Scudai River scheme should be
deemed as a permanent supply source and that the Johore River scheme should
be constructed as quickly as possible to produce, at least initially, an additional

60 Mgal/day. It was estimated this project would cover Singapore's needs until 1982.

Binnie & Partners recommended that additional treatment capacity was built in Johor. They reached this conclusion after considering the plant capacity and the related saved costs, which would be minimal. Moreover, a centralized treatment system would result in a significant loss in the flexibility of delivery mains, the laying of some additional pipelines and the hazards of operating raw and treated water pipelines in an interconnected system.

The 1972 Water Master Plan

As an independent nation, the importance of an autonomous source of water supply was as critical in 1965 as it is at present. The comments made in 1972 by the PUB Water Department's Superintending Engineer Sung, represent the overall thinking about this issue:

> In the former days, the water planner had a choice to look for a high quality water source for collection from across the Causeway and also an opportunity to perform economic appraisals on alternative sources and schemes. However, immediately after its independence in 1965, water supply planning criterion, likewise, has to undergo a radical change. The immediate task facing the water planner is logically to plan and develop all available internal resources as quickly as possible and as much as practicable. The desirability to achieve self-sufficiency in this vital commodity is obvious.
>
> (Sung, 1972: 18)

In 1971, a Water Planning Unit was set up under the Prime Minister's Office to study the scope and feasibility of new conventional sources, such as unprotected catchments, and unconventional sources, namely water reuse and desalination. The Water Planning Unit consisted of three officers: Lee Yong Siang, Chou Tai Choong and Tan Gee Paw. The Planning Unit reported directly to the PUB Steering Committee on Water Resources chaired by the then Chairman of PUB and Minister of Education, Lim Kim San, and was comprised of three other officers, Khong Kit Soon, the acting PUB General Manager; Lee Ek Tieng, Acting Permanent Secretary at the Ministry of Health; Lee Yong Siang, Head of the Water Planning Unit. The outcome of the study was the first Water Master Plan in 1972, which outlined the strategies for local water resources in Singapore, including water from local catchments, recycled wastewater (NEWater), and desalinated water, to ensure a diversified and adequate supply to meet future projected requirements (Tan *et al.*, 2009).

Initially, the Water Planning Unit sought the expertise of Tahal consultants from Israel to prepare the Water Master Plan. At that time, Singapore engaged a number of experts from the United Nations (UN) to prepare long-term plans, including the 1971 Concept Plan and water resources studies. Tahal consultants were brought on board because they were perceived to have the expertise in maximizing water resources given their geographical constraints. However, the Water Planning Unit soon realized that the hydrological conditions in Israel and Singapore varied too much for the Israeli consultants to be able to make any meaningful contribution. For example, while the Israelis were experts on drip irrigation techniques to minimize water usage for irrigation in dry areas, Singapore experienced heavy rainfall almost all year round and the Israel experience was thus not relevant. At that time, the Israeli consultants had even suggested that Singapore should consider putting a layer of oil or emulsion on the reservoir to prevent evaporation (Falle, 1971; personal interview with previous Prime Minister, Lee Kuan Yew, 11–12 February 2009). Thus, the planners quickly realized that the Israelis did not have the practical experience to deal with water problems in tropical regions and discontinued their services, embarking on the 1972 Water Master Plan on their own (personal interview with previous Prime Minister, Lee Kuan Yew, 11–12 February 2009).

Similar to the 1962 plan, the 1972 Water Master Plan had a 20-year planning horizon, with water demand projections being estimated up to 1990 (PUB, 2010, personal communication). In contrast to earlier projections, in 1972 it was acknowledged that a single large water-consuming industrial complex could upset demand forecast by a wide margin at short notice (Sung, 1972).

Water supply options

The concept of unconventional water sources (reuse wastewater and desalinated water) was introduced in the 1972 Water Master Plan. The overall water supply development strategy was to proceed with surface water schemes as fast as possible to increase supply and to keep under sight unconventional supply sources for projects to be implemented when it became technically feasible or necessary.

In terms of surface water sources, the objective was to extend the water catchment area from the central protected catchment, which covered just 11 per cent of the island's area, to the western and eastern regions, covering 75 per cent of the entire area. The shaded areas in Figure 1.2 show the extent of the proposed water catchment area. Proposed initiatives included the following schemes: Bedok Reservoir, Kranji-Pandan, Western Catchment and North-eastern Singapore. At that time, the Kranji-Pandan scheme was already being implemented. Yet, planners acknowledged that besides the protected central catchment area, the other water catchments were heavily polluted and required the implementation of extensive pollution control measures before the reservoir schemes would be realized.

Sung noted that, in 1972, the existing developed area in Singapore represented some 29 per cent of the island's area, and, thus, that water supply schemes should be planned to co-exist with other land users by 'unconventional approaches and thinking' (Sung, 1972: 220). In addition to urban storm water runoff schemes, some of the unconventional surface water options included estuary-barrage schemes to dam up rivers and turn them into freshwater reservoirs. Figure 1.2 shows the existing and proposed reservoirs and water catchments in 1972.

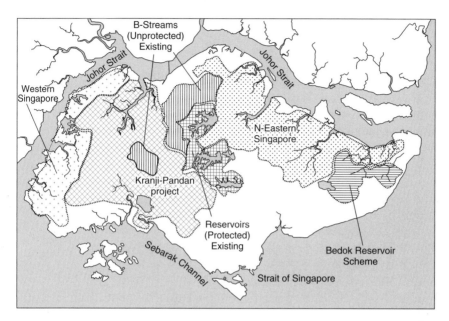

Figure 1.2 Existing and proposed reservoirs and water catchments in 1972 (Sung, 1972).

Regarding groundwater, water planners reviewed and updated the 1950 groundwater study carried out by White. Contrary to the 1950 preliminary findings, it was found that the groundwater potential in the eastern part of Singapore was very limited because of poor prospects of natural recharge from the ground surface. The soil was virtually impermeable and the lithology of the sediments forming the alluvial basin was not favourable to this end. Cloud seeding was also unsuccessful, although the plan considered it as a feasible alternative for the future.

It was then that the role of recycled water started to be more thoroughly considered:

> There is not the slightest doubt that in time to come, reclamation of sewage effluent for rational use will become very much an economic necessity. The important role it will assume as a normal part of Singapore's future urban water cycle can well be imagined . . . the direct use of treated effluent as a source of potable water supply is not advocated. Apart from its psychological objection, there remains always the real health danger of a breakthrough during treatment even for a very brief interval . . . a major part of these objections might be removed if the treated effluent is first diluted in a large impounding reservoir affording long retention period to undergo [a] self-purifying process prior to withdrawal for treatment, needs to be fully explored.
>
> (Sung, 1972: 23)

The planners also acknowledged the unlimited potential of seawater and the fact that desalination had the benefit of not being subject to unpredictable weather conditions. Desalination was thus considered as a feasible and dependable alternative to be initially explored during drought periods. Nevertheless, as the cost of desalination was still very high, it would be pursued only after natural freshwater resources were fully exploited and developed.

Decision-making to increase water supply: the whys and hows

Throughout the years, when planning for water resources development, the thinking of water security has been a permanent consideration for the city-state's leadership. For example, during a debate on the need to increase water prices to find and fund new water supply schemes in 1970, the then Minister of Finance, Goh Keng Swee, explained that Singapore had been developing local water supply schemes so that it could be 'dependent as little as possible on outside supplies'. At the same time, since the setting up of water supply and storage facilities in Singapore was more expensive than importing water from Johor, water engineers would start working on the most economical schemes before going on to more expensive ones. He added that 'this [was] the price [the country had] to pay for not giving unnecessary hostages to fortune' (Goh, 1970: column 423).

As mentioned before, the government policy for the expansion of water supply sources was to develop all available water resources (Lee, 1986). Up to 1986, these included water from local catchments. It was only after all surface water resources had been exhausted in the late 1980s that Singapore seriously looked into developing unconventional water supplies.

1960–1986: developing surface water sources under the 1972 Water Master Plan

Between the 1960s and the 1970s, the water supply development schemes implemented were the continuation of early plans set out by the British. Singapore started by expanding two of the three existing reservoirs in the central catchment area. In 1969, the Seletar Reservoir was expanded 35-fold to create the Upper Seletar Reservoir and in 1975 Peirce Reservoir was enlarged ten-fold creating the Upper Peirce Reservoir. The Scudai and Johor River schemes were also developed and set out under the 1961 and 1962 Water Agreement respectively to import water from Malaysia. These two initiatives were completed with the 30 Mgal/day Scudai waterworks (finalized in 1964) and with the first 30 Mgal/day of Johor River waterworks (completed in 1967), just in time to meet Singapore's increasing water demands.

Subsequently, estuarine barrages were created in unprotected water catchments to form reservoirs. These were built by damming the mouths of the western rivers and pumping out the brackish water so that the rivers could be used to store freshwater. In this way, the Kranji-Pandan scheme was finalized in 1975 with the completion of Kranji and Pandan reservoirs. By 1981, the Western Catchment scheme, comprising the Murai, Poyan, Sarimbun and Tengeh reservoirs, was also completed (PUB, 2002b).

Once the aforementioned reservoirs were built, Singapore turned to unprotected water catchments in urban areas to collect water from densely populated new towns. The first of these urban water catchments was the Sungei Seletar-Bedok scheme, completed in 1986, along with the creation of the Bedok and Lower Seletar reservoirs and water treatment waterworks at Bedok (PUB, 2002b).

In 1971, the idea of developing a reservoir at Bedok by reclaiming land and using a sand quarry was raised (Barker, 1971). However, while the water storage capacity at Bedok was potentially very large, the catchment was quite small and the resultant quantity of water that could be tapped on was not particularly significant. In addition, the area was being developed as a new HDB town, meaning Bedok would be an urban water catchment and thus runoff water quality was expected to be more polluted, especially during dry weather (Barker, 1976). Consequently, the PUB had to study both the related problems of water quantity and quality.

To address quantity constraints, the PUB proposed increasing the runoff that could flow to Bedok Reservoir by indirectly increasing the water catchment area. This was done through runoff collection from the Lower Seletar water catchment where a reservoir was also developed (north-east of Bedok new town), and sending it to Bedok Reservoir. Moreover, in solving quality problems, interagency coordination played a big role. The ENV did its part to extend the sewerage network to ensure that all used water was collected and treated. Since the Bedok site was already earmarked under the 1971 Concept Plan as a potential water catchment area, the Urban Redevelopment Authority (URA), which oversees land use planning in Singapore, zoned land use and safeguarded it against polluting developments. The HDB excavated the sand that it required for its future projects and stockpiled it elsewhere so that Bedok Reservoir could be completed in time to meet increasing water demands (Tan *et al.*, 2009). The Sungei Seletar-Bedok scheme was finally completed in 1986.

1960–1986: first trial with water recycling

Industrialization in Singapore began in the early 1960s. Part of this effort was the development of the Jurong Industrial Estate, an area of about 14,000 ha located on the south-west coast that was to be developed as an industrial satellite town. Besides the question of whether the estate would be economically successful, there was also the question of how to meet industrial demand. To this end, and as the main driver of the industrialization programme, the EDB initiated a study on how to satisfy Jurong's industrial water requirements for which the Board engaged consultants from the UN Economic Commission for Asia and the Far East (ECAFE). The group found out that treated effluent from the Ulu Pandan wastewater treatment plant could be further treated to meet the industry's non-potable water demand. The Ulu Pandan wastewater treatment plant was chosen because its location in the west made it geographically close to the industrial estate and the facility mostly treated domestic wastewater (ECAFE, 1964).

Jurong Industrial Waterworks (JIWW) was completed by the EDB in 1966 and introduced as an inexpensive source of lower-quality water for industries that did not require potable water. In 1971, and in collaboration with the HDB, the ENV took over the waterworks and in addition to supplying water to the industries at Jurong Industrial Estate, embarked on a trial scheme to supply industrial water for toilet flushing to flats at Taman Jurong, Pandan Garden and Teban Garden in the western part of Singapore. However, the lower water quality resulted in high maintenance and replacement costs, which negatively affected the viability of the scheme. It was eventually discontinued in 1990 (Tan *et al.*, 2009).

During this period, Singapore also carried out its first trials with more advanced wastewater treatment technology. Industrial water was created by passing treated effluent from the Ulu Pandan wastewater treatment plant through the conventional treatment process of coagulation, flocculation, clarification, sand filtration and aeration. However, as Singapore was interested in recycling water that was good enough to drink, an advanced pilot water reclamation plant was established in 1974. Interestingly, the pilot plant was set up by the ENV's Sewerage Department and not the PUB's Water Department. The secondary treated effluent underwent reverse osmosis and other advanced treatment processes of ion exchange, electro-dialysis and ammonia stripping to produce water complying with drinking standards. The water even met the World Health Organization (WHO) guidelines for drinking water. However, membranes were expensive, making the water reclamation process not cost-effective. Additionally, membrane technology was unreliable given that membrane fouling posed a significant challenge that required frequent cleaning. After 14 months of operation, the advanced water reclamation demo plant was shut down in December 1976.

1960–1986: water supply strategy for the first two to three decades

The 1972 Water Master Plan took into consideration the importance of reducing dependency on imported water for the country; the lack of water storage capacity to retain and make use of the large amounts of rainfall; the need to construct more reservoirs in light of limited options due to incompatible land usage; and that, even if the whole island was a large reservoir, water supply would still not be sufficient to meet the rising demands. The Master Plan thus included action plans to develop more cost-effective water supply schemes first; extend the water catchment area from 11 per cent to approximately 75 per cent of the island's area; implement pollution control measures to ensure that the water catchment area could be increased; and look into unconventional sources such as desalinated and recycled water after maximizing surface water sources. Groundwater was not an option because of the lack of it.

The implementation of the different alternatives to develop water supply projects in Singapore was a complicated matrix that considered aspects such as quantity, quality, cost, reliability and security, to mention only some of them. Surface reservoir schemes were the most obvious sources to tap. Local water catchments and reservoirs would not only add to Singapore's water supply capacity, but also had the strategic advantage of increasing the water storage capacity. Even when these schemes presented the most cost-effective alternative, it was still an expensive operation.

As a then developing nation, Singapore faced the same problem that many developing countries still face today: it lacked the heavy capital investments necessary to develop water infrastructure. Thus, early reservoir projects were all funded by loans from international organizations such as the World Bank (expansion of the Seletar Reservoir) (PUB, 1970), the Commonwealth Development Corporation (expansion of the Peirce Reservoir) and the Asian Development Bank (ADB) (Kranji-Pandan and Western Catchment schemes) (PUB, 1970, 1975). Later on, Singapore also borrowed from the Bank of America (in a co-financing scheme with the ADB) for the Western Catchment scheme (PUB, 1985a). These loans were all successfully paid back in accordance with the stringent conditions stipulated by the international development banks and these rules perhaps went some way to instill financial discipline in the relatively young organization.

The development of water supply sources can be best summarized with the timeline in Figure 1.3.

Whereas the 1960–1970 period was devoted to the fast development of projects to import water and meet increasing demand, the following years centred on building up local water supply sources to allow Singapore to be less dependent on Malaysia. In 1986, once the Sungei Seletar-Bedok Reservoir scheme was completed, the PUB could then focus on urbanized catchments, such as Marina, as well as on technology that could reduce the costs of desalination and recycled water, potentially increasing local water supply.

1986–1994: further exploring unconventional water sources

The Sungei Seletar-Bedok water supply scheme completed in 1986 was thought to be the last of surface reservoir projects (PUB, 1987). Subsequently, Singapore started looking more seriously at other alternatives, which included new sources of imported water and unconventional water sources such as desalinated and large-scale recycled water to later on reconsider undertaking new reservoir schemes.

That same year, in response to questions on future water supply plans, the then Minister of State for Defense and Minister of State for Trade and Industry, Lee Hsien Loong, explained that after the Sungei Seletar-Bedok water scheme, the PUB would have to develop water resources outside of Singapore, including extending the water treatment capacity at both the Scudai and Johor rivers. However, since desalinated water was '10 times as expensive as processing river and reservoir water', Lee commented that '[i]f people need water so badly that they are prepared to pay more than 10 times as much as they now do for water, then it will be done' (Lee, 1985: column 1608). For this, the government had to carefully tread the line between being confident of having sufficient water resources, and still remind the public that water conservation was extremely important because

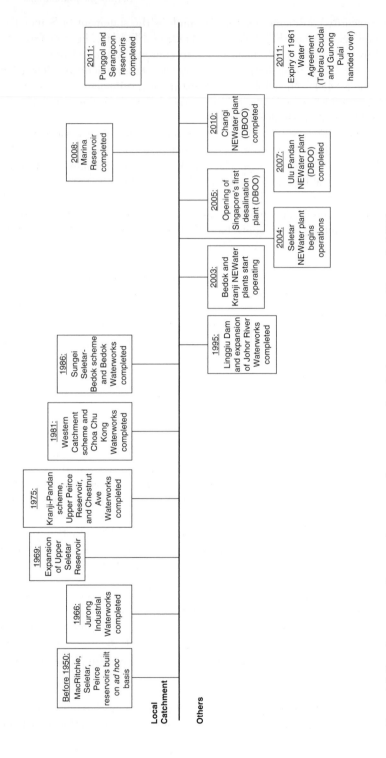

Local Catchment

Before 1950: MacRitchie, Seletar, Peirce reservoirs built on *ad hoc* basis

1966: Jurong Industrial Waterworks completed

1969: Expansion of Upper Seletar Reservoir

1975: Kranji-Pandan scheme, Upper Peirce Reservoir, and Chestnut Ave Waterworks completed

1981: Western Catchment scheme and Choa Chu Kong Waterworks completed

1986: Sungei Seletar-Bedok scheme and Bedok Waterworks completed

2008: Marina Reservoir completed

2011: Punggol and Serangoon reservoirs completed

Others

1995: Linggiu Dam and expansion of Johor River Waterworks completed

2003: Bedok and Kranji NEWater plants start operating

2004: Seletar NEWater plant begins operations

2005: Opening of Singapore's first desalination plant (DBOO)

2007: Ulu Pandan NEWater plant (DBOO) completed

2010: Changi NEWater plant (DBOO) completed

2011: Expiry of 1961 Water Agreement (Tebrau Scudai and Gunong Pulai handed over)

Figure 1.3 Developments in Singapore's water supply sources timeline (information from Tan *et al.*, 2009).

of water security considerations. One example is the statement of Senior Parliamentary Secretary to the Minister for Trade and Industry, Eugene Yap, who stressed in Parliament how crucial it was to conserve water and protect water resources 'because it is likely that we will not be self-sufficient in water' (Yap, 1988: column 1484). The public could construe such statement as a lack of governmental confidence in Singapore's water supply for which the State had to be very careful in making similar statements.

During the 1980s and 1990s, there were two main arguments against desalinated water. First, desalination had not been implemented on a massive scale and thus the technology was not proven on a large scale, and second, desalination was very energy intensive, and thus, very expensive. The then Minister of State for Defense and Minister of State for Trade and Industry, Lee Hsien Loong, gave the example of Hong Kong, where a desalination plant was constructed in response to China's threats to cut off their water supply. Nonetheless, the plant had to be closed in the end because it was simply too expensive to operate. Even so, the Singapore government was confident that water supply was not an issue; and it emphasized that it was just a matter of cost (Lee, 1986). In 1995, the then Minister of Trade and Industry, Yeo Cheow Tong, made a statement reassuring people that:

> [t]here is no question that we will solve the water problem. Even if we have to desalt every drop of water that we drink or cook with, as countries do in the Middle East, we can do so. It will cost us a lot of money, but it will not bankrupt us. Fortunately, we do not have to do this, as we still have other sources available.
>
> (Yeo, 1995: column 198)

There were also suggestions to use recycled water for non-potable uses such as domestic washing and toilet flushing. At that time, while it was technically possible to produce recycled water that met drinking water standards, the 1974 pilot plant study had shown that the process was significantly costly and technology unreliable. Thus, recycled water plans mostly referred to using recycled water of non-drinking quality. This brought with it a host of technical and cost considerations. For example, a separate (and very pricey) reticulation system would be required for this lower grade water, and there was the risk of cross-contamination, not to mention aesthetic issues resulting from the smell emanating from water that would not be completely clean.

In the late 1980s, the government also began studying Marina Bay as a secondary source of non-potable water (Yap, 1988). The idea was to meet non-potable water demand and possibly also use it as a potable source during times of emergency. Yet, this initiative was only remotely considered.

Regarding further alternatives to import water to Singapore, Kog describes it as follows:

> In 1987, then Prime Minister Lee Kuan Yew announced that Singapore was looking into the possibility of tapping water from Indonesia (Business Times

7–8 Oct 1989). Following this, an agreement 'on economic co-operation in the framework of the development of the Riau Province' was signed on 28 Aug 1990. Under this agreement, Singapore and Indonesia agreed to co-operate on the sourcing, supply and distribution of water to Singapore. This agreement also includes co-operation over trade, tourism, investment, infra-structural and spatial development, industry, capital and banking (Govern-ment Gazette, Treatises Supplement No.1, 1990). In 1991, a Water Agree-ment signed with the Indonesian Government provides for the supply of 1,000 million gallons of water a day from sources in the Province of Riau in Indonesia.

Also planned is a project to tap the water resources of the Sungei Kampar catchment in West Sumatra (ST, 30 Jan 1993). The Sungei Kampar project is described as a project to provide water to the Riau province. This suggests that water for Singapore would have its place as part of a much larger plan to build up the regional economy. The Bintan project marks the beginning of a new era in the development of new water projects. While previously the PUB has managed the construction and development of water resource projects alone, the Bintan scheme will be built and managed by PUB's sub-sidiary company, Singapore Utilities International (SUI). Two joint venture companies were created in 1992 for the purpose of developing a supply of water from Bintan and to supply water to Bintan and the neighbouring Riau islands (PUB, 1992). However, the pace of the progress for these develop-ments has been very slow and it has come to a halt because of the political uncertainty in Indonesia after the financial crisis.

(Kog, 2001: 20)

On the Malaysian front, plans were also made to increase water imports. While the 1962 Water Agreement allowed Singapore to draw up to 250 Mgal/day of water from the Johor River, Johor River waterworks were designed with a capac-ity of 120 Mgal/day, which was the amount the PUB could reliably draw from the river based on its flow. Plans were thus started to develop the necessary infra-structure to fully draw the amount of water allowed under the 1962 Water Agree-ment. Moreover, after negotiations with Malaysia, the 1990 Water Agreement was signed as a supplement to the 1962 Water Agreement, both to expire in 2061 (see Agreement between the Government of the State of Johor and the Public Utilities Board of the Republic of Singapore, signed in Johor on 24 November 1990).

The 1990 agreement allowed the PUB to construct a large dam upstream of the Johor River waterworks to collect and store runoff from the upper reaches of the Johor River catchment. Water could thus be released to regulate the river flow and allow higher volumes to be drawn at the intake point of the waterworks. This document also included provisions for water sales beyond the 250 Mgal/day stipulated under the 1962 Water Agreement. The price of additional intakes would be calculated based on: the weighted average of Johor's water tariffs plus 50 per cent of the surplus from the sale of this water by the PUB to its consumers after deducting Johor's price and the PUB's cost of distribution, or 115 per cent of the weighted average of Johor's water tariffs, whichever was higher. With this agreement in place, the Linggiu Dam was completed in 1993 and the Johor River waterworks were expanded to 160 Mgal/day that year and subsequently to 250 Mgal/day in 2001. This arrangement was thus crucial in enabling Singapore to draw its 250 Mgal/day entitlement from the Johor River. So far, neither country has made use of the clause on additional water supplies beyond 250 Mgal/day (PUB, 2010, personal communication).

This augmented water supply allowed Singapore to meet its growing water demands at the same time that it gave it time to explore further unconventional water sources that were still too expensive and technically unfeasible.

1995–2011: exploring unconventional water sources

Singapore began to explore unconventional water sources in the mid-1990s. In 1996, consultants were appointed to carry out site feasibility and engineering studies on desalination to address issues such as site location, desalination processes and the cost of desalinated water. The plan was for the first desalination plant to have a capacity of 30 Mgal/day and be completed in 2003 (PUB, 1996). In 1998, the engineering and site feasibility studies were completed. It was decided that the plant would be built on reclaimed land at Tuas and that the first phase of the desalination plant, with a capacity of 30 Mgal/day, would be constructed based on the dual-purpose multi-stage flash distillation process with an adjacent power plant by 2005.

The PUB also announced that it was studying the possibility of further increasing local sources by developing suitable marginal catchments including schemes to tap storm runoff from new housing estates in Hougang, Punggol and Sengkang. Rainwater would be collected and treated to meet drinking water standards instead of simply being drained (for flood control) and sent out to the sea. The PUB explained that these schemes would be implemented along the development of drainage systems in the new towns. The initiative's cost was estimated at $170 million and would increase Singapore's total catchment area by 5,500 ha (PUB, 2010). At the same time, the PUB and ENV embarked on a joint assessment on the feasibility of water reclamation using secondary treated sewage effluent. The $14 million study involved the construction of a demonstration plant with a 10,000 m^3/day capacity to use advanced membrane technology to treat sewage effluent that would meet the internationally accepted WHO Drinking Water Standards (PUB, 1998b).

In 1999, it was decided that the desalination plant should be built by the private sector from which the PUB would purchase the water. It was also agreed that a smaller 10 Mgal/day desalination plant would be owned and operated by the government (PUB, 1999). Bidders were free to choose from a range of available desalination processes including multi-effect distillation, multi-stage flash distillation, reverse osmosis or hybrid systems. Reverse osmosis was a new process that had become available only after advancements in membrane technology in the 1990s. Before membrane treatment of saltwater was developed, distillation was the main technology used. Different distillation variations relied on large amounts of energy to produce heat or the required pressure conditions to evaporate water that would then condense on a cooler surface, a process that made distillation very expensive. SingSpring Pte Ltd, a subsidiary of Hyflux Pte Ltd, won the bid and Singapore's first 30 Mgal/day reverse osmosis desalination plant was opened in 2005.

On the recycled water front, a demonstration plant was built at Bedok in 2000 and an international panel comprising both local and foreign experts was formed to provide independent advice on the study. Unlike the recycled water trial in 1974, technology had advanced to a point where high-grade reclaimed water could be produced through a multi-barrier treatment process that comprises conventional used water treatment, micro/ultrafiltration, reverse osmosis and, finally, ultraviolet disinfection.

After two years of careful analyzis, recycled water was found to be safe for drinking and, in 2002, the plan for tapping on recycled water as another water resource was set in motion: new plants were built and, equally importantly, a communication plan was put in place. A fundamental part of this outreach effort was not necessarily to focus on the technology employed, but to educate the public that this recycled water was safe for drinking. In order to try to change the overall negative popular impression towards recycled water, recycled wastewater was renamed as 'NEWater', wastewater treatment plants renamed as 'water reclamation plants', and wastewater as 'used water'. More importantly, the new terms were part of a strategy that aimed at achieving a mindset shift, stressing the new approach to water management by communicating to the public the need to look at water as a renewable resource that could be used over and over again. At present, NEWater is supplied both for direct non-potable use (DNPU) to commercial and manufacturing processes that require water for cooling, and for indirect potable use (IPU) by introducing water into reservoirs for subsequent retreatment at the several waterworks for drinking purposes (for an extensive account on NEWater, see Tan et al., 2009).

Similar to desalinated water, NEWater was also opened up for private sector participation. The first three plants built in 2003 in Bedok, Kranji and Seletar were owned and operated by the PUB. However, the fourth 32 Mgal/day Ulu Pandan NEWater plant was built under a Design-Build-Own-Operate (DBOO) model with the private sector. The main reason behind the DBOO model was to develop a water industry that would provide services at a specific level of quality at a cost-effective price, as well as to encourage greater efficiency and innovation in the water sector (Tan et al., 2009). The Ulu Pandan NEWater plant was completed in 2007. Finally, the fifth 50 Mgal/day Changi NEWater plant, also built under a DBOO approach, was completed in 2010.

In 2008, the Marina Reservoir was completed. Known as the Reservoir-in-the-City, it is located right in the heart of the central business district, becoming the most urbanized and largest catchment in the city-state. The reservoir has been created through the construction of a barrage across the Marina Channel, at the confluence of five rivers (including the historic Singapore River) to collect and store water from the most densely populated areas. In 2011, two additional reservoirs at Punggol and Serangoon were opened.

Further thoughts

As part of Singapore's overall development strategy, a multiplicity of issues have been taken into consideration along the years. These have included an overall vision; clear objectives; long-term planning; effective legislation; institutional coordination amongst government ministries and agencies, and between government and the private sector; and the general public's cooperation, to mention only some. Aware of the importance clean and reliable water supplies have for the security of the city-state, protection of water catchments as well as development and diversification of sources have historically played a crucial role in the overall development of the city-state.

Looking back, when pondering how Singapore succeeded against all odds in achieving overall development in spite of its limited size and scarce natural resources (mostly water), one should acknowledge the part the country's leadership and its vision have played in this process. Early on after independence, PAP leaders realized that the economy was a priority issue, which compelled the move away from entrepôt trade and into industrialization geared to export markets. The PAP was also aware that nation-building could succeed only if the population of Singapore was committed to the development of their city-state. For this support to exist, people would have to see their aspirations fulfilled through economic improvements that would have to be reflected in their quality of life (Quah, 2010).

It was thus the importance of economic and social development that made provision of services so crucial in the early years of nationhood. Singapore's ability to establish utilities and provide basic services rapidly showed the efficiency of the newly independent country. This was vital for the population to trust the government and to create a virtuous financial circle where taxes collected by these investments could in turn be used for additional infrastructure.

Regarding planning instruments, two documents have been essential for Singapore, namely the 1971 Concept Plan and the Master Plan. As Tan *et al.* (2009) describe, government agencies worked together during the preparation of the Concept Plan to outline a long-term land-use based vision for the development of the city-state and the improvement of the population's quality of life. The Concept Plan presents broad, long-term strategies, but it is the Master Plan that has guided Singapore's development for more than a decade, translating far-reaching and visionary strategies into such detailed plans that specify the permissible land use and population density for each plot of land.

Forward planning has been critical for the development of Singapore's infrastructure. Even though some plans may not have been immediately implementable due to resource constraints, their early consideration has made possible their posterior tracking and implementation once it has become feasible to do so. For instance, the 1972 Water Master Plan proved to be an innovative document as it considered not only those surface water sources that were to be developed over the next 20 years, but also unconventional water sources that would be tapped on much later.

Water planning requires coordination and, as such, in 2001, important institutional reorganization took place as the Drainage and Sewerage departments were combined with the PUB's Water Department. Such a merger led to the creation of a new PUB in order to have all government departments responsible for the different aspects of the water cycle under the same organization. Tan Gee Paw, one of the members of the Water Planning Unit responsible for the first Water Master Plan, returned to the PUB as its Chairman. Upon his return, he noted that the Water Master Plan had not yet been reviewed, nor had it been discussed what should be Singapore's plans after the water agreements expired (Singapore National Archives, 2007). The Chairman thus started the process to develop a new long-term plan for the development of water resources beyond the expiry dates of the two water agreements in 2011 and 2061. Since all conventional surface water sources under the 1972 Water Master Plan had been developed, new alternatives had to be looked into, namely the amount of rainfall that could be collected, the volumes that could be recycled and how much could be desalinated.

Looking for answers to these and other concerns, as well as looking to go past foreseeable challenges, sensible policies and programmes have been established along the years. In its rightful ambition to achieve water self-sufficiency and aware that the lack of clean and reliable water resources could hamper its own development, Singapore has developed a pragmatic vision, long-term planning and action frameworks that have allowed it to move from vulnerability into sustainability. On the other hand, while the government has always stressed that it is of utmost important to reduce dependence on outside supplies and diversify domestic water sources, it has never revealed any quantitative goals or timelines for self-sufficiency.

In Singapore, water development goals have been achieved in the larger context of social and economic development. In spite of many challenges and difficulties, political leadership has worked together with the several governmental agencies and the general population to put the city-state on the track to achieve security as well as sustainable development. When asked about the factors that have contributed to Singapore's success in turning water from a vulnerability to a strength, previous Prime Minister Lee Kuan Yew noted:

> it was critical circumstances, determination to succeed, comprehensive planning and the technology . . . The same process can be repeated by any country. But you must have the determination, the discipline, the administrative capability and its implementation. And you keep on looking for new technology.
>
> (personal interview with Lee Kuan Yew, 11–12 February 2009)

It has thus been a combination of political will and leadership, clear vision, long-term planning, pragmatic decision-making, adequate policy implementation and coordination between government agencies and public, private, civic and civil sectors that has allowed Singapore to embark on the path towards water security and sustainable development.

2   Water and urban development

Introduction

Singapore's foundations for its journey towards sustainable development were laid through sound land, water, infrastructure and environmental policies as early as the 1960s, when the term 'sustainable development' was still not even in vogue. Unlike the earlier newly industrializing economies of East Asia like South Korea, Taiwan and Hong Kong who adopted the 'industrialization model first and consequences later' approach, and as a recently independent nation, Singapore decided from the very beginning to implement sound environmental policies and conserve its natural resources in tandem with planned economic growth and development (Ooi, 2005).

The quest for water self-sufficiency that has sought to tap on available domestic resources has had an important impact on the way urban development has been planned. As such, it has shaped how space is used in the land-constrained island. This chapter discusses how water has been incorporated as a key factor in Singapore's urban development process in the context of rapid urbanization and industrialization, an experience that stands as a valuable case study for sustainable urban development.

Even though the literature on the role of the environment in sustainable urban development is vast, water has only been addressed as a minor component. In the case of Singapore, however, the unique status given to water warrants that a much larger role is awarded to it mainly for two reasons. First, because sustainable urban development has required proactive and holistic planning to overcome socio-political and economic barriers that would have made it impossible to pursue otherwise; and second, because the city-state has emphatically proven how the conservation of scarce natural resources can be improvised in synchronization with planned urban development.

In Singapore, the fundamental importance of sustainable development has been acknowledged by the leadership of the city-state as a matter of necessity rather than choice. The relentless urge to improve in all aspects of public and private life has resulted in pragmatic policies that have followed a clear vision, long-term planning and forward-looking strategies. They have been implemented in response to the changing needs and goals of the city-state and have set the foundations for its overall development and growth.

Lee Hsien Loong, current Prime Minister of Singapore, summarizes the meaning sustainable development has for the city-state and the achievements made in this regards since independence:

> Sustainable development means achieving the twin goals of growing the economy and protecting the environment, in a balanced way. Singapore has practiced sustainable development even before the term was coined. We pursue growth in order to have the means to improve our lives. We also safeguard our living and natural environment, because we do not want our material well-being to come at the expense of our public health or overall quality of life.
>
> Singapore is a small island with finite space, limited water supplies and no natural resources. Yet, we have overcome our constraints, grown and developed into a modern city. Through imaginative city design, careful planning and judicious land use, we have housed close to five million people in a clean and green city, with one of the best urban environments in the world . . .
>
> . . . Sustainable development demands long term attention and effort. Some measures will incur disproportionate costs and impair our competitiveness. We have to adopt a pragmatic approach, find the most cost-effective solutions and pace the implementation appropriately so that we do not hurt our economy. We should also invest in capability building and R&D, to take advantage of new technologies that facilitate sustainable development.
>
> This issue concerns not just one or two ministries, but the whole country. Hence we will tackle it using a whole-of-government approach. The people and private sectors also have to work with the public service on this important venture.
>
> (Lee, 2009: 6–7)

Lee's words reflect Singapore's particularities whilst all the while touching on the political, economic and social strands sustainable development seeks to bring together. The journey the city-state has embarked on, its urban development process and the role water resources have played at different times in history are analyzed in this chapter. Even when direct comparisons between the city-state and other rapidly industrializing cities prove difficult given its unique position as an island nation, Singapore did face many specific and shared challenges. Difficulties have been even more constraining for the city-state as there is no hinterland to rely on to bring in natural resources and therefore the adverse impact growth has on the environment has always been absorbed by the tiny island itself.

The next few pages thus assess the 'Singapore way' towards sustainable development. That is, the implementation of urban development policies in spite of, because of and thanks to land, water and natural resources limitations. This 'way' also draws attention to how environmental policies have been an integral part of the overall development strategies the city-state has pursued (Ooi, 2005).

Foundations for a sustainable urban development

Urban planning is not a new concept in Singapore. Within four years of his arrival in 1819, Sir Stamford Raffles instructed a master scheme for the development of the city, a document completed in 1823 and later called the Raffles Plan. Drawn by Lieutenant Jackson, this plan intended to exploit the natural advantages of settling in the site, developing the town in such a way that natural site opportunities were seized to supply water, build infrastructure and defence (Waller, 2001).

As a port-city, water was, and remains, fundamental not only to cover the needs of local residents but also to replenish supplies for the ships. In those days, Singapore used to import low-cost goods, which after repackaging were shipped again to various destinations. Since it was easier to handle the goods in the sheltered waters of the island's main waterway than along the exposed foreshore, the Singapore River soon became the principal point of entry to the town as well as a trade artery (Dobbs, 2003). By the end of the nineteenth century, economic progress emerging from the growth of the port businesses and a rise in population brought not only wealth to Singapore but also problems associated with a busy metropolis. Housing constraints and land and river pollution soon emerged as urgent matters to be taken care of. Yet, it took many more decades to draft a concrete programme.

While infrastructure was developed on an ad hoc basis throughout the rest of the nineteenth century and first half of the twentieth century, it was not until 1955 that the Singapore Improvement Trust (initially established in 1927 to build new housing estates) began discussing the need to elaborate a Master Plan that would respond to the then economic and social situations. The 1955 Master Plan was the result of extensive surveys on land use, traffic, car parks, etc., and on assessments on population growth and employment trends. It was supposed to cover the 1953–1972 period, but it was approved only in 1958, when there was already a growing need for a comprehensive management of both land and environment. The plan, with a 20-year horizon, had the main objective to alleviate the acute housing shortage and redevelop and improve the living conditions in the central area by reducing about one-sixth of the resident population and reallocating it to the surrounding urban areas. It also introduced the concept of, and actions for, a planned environment with a green belt surrounding the city area that would prevent urban growth and the development of three satellite towns to decentralize activities, namely Jurong, Woodlands and Yio Chu Kang (Yeung, 1973; Waller, 2001). The scheme made major emphasis on the rational use of land through zoning demarcations for all urban land uses (Dix, 1959). Even though the plan laid down the basis for comprehensive urban planning, this exercise exposed one important caveat, which was the tendency to treat the scheme as an end in itself rather than as a part of a wider planning process (Teo, 1992).

In 1959, Singapore gained self-government with the People's Action Party (PAP) winning the parliamentary elections. Soon after would start the difficult, painstaking success story of the city-state.

After the elections, it became clear for the PAP that a new plan was required to reflect the changing socio-economic and political conditions. The party thus set out a five-year plan to improve the lives of the entire population as well as to remove the rural–urban disparity in terms of access to life amenities and basic services (PAP, 1959).[1] The resulting rural development programmes set water and electricity delivery to the rural population as the main goals to be reached. To fulfil the objective stated for the five-year plan, but also for the long-term economic and social development of the city-state, it proved adequate and much more pragmatic to plan for concentrated flats rather than to allow sporadic rural slums to emerge. Furthermore, bureaucratic red tape was so prevalent in those days that the provision of facilities in a housing complex was to be significantly more straightforward, instead of having to give answers to hundreds of individual requests. Always following the practical policy design and implementation that came to characterize the ruling party, PAP leaders emphatically explained to the electorate the need for low-cost housing units in a land-scarce place like Singapore. Consequently, special emphasis was given to public housing (Ong, 1959b).

Challenged by a growing population and unemployment rate, the creation of employment opportunities became the newly elected government's overriding objective. In 1960, it was evident that a process of immediate urban renewal had to be carried out to support and encourage economic growth. This course of action included clearing slums, providing alternative housing for those displaced, improving the living environment and revitalizing the city centre. At that time, an estimated quarter of a million people (approximately 60 per cent of the total population at that time) were living in city slums, which implied that between 1961 and 1970 some 147,000 new dwellings would be required to house them (Quah, 1983). Such a situation called for the pragmatic formulation and implementation of a wide range of social and economic policies, a key element that has remained decisive in shaping Singapore's success.

With the changing conditions, two important ordinances were enacted to take over the duties carried out by the colonial Singapore Improvement Trust. One of these statutes established the Housing & Development Board (HDB) in February 1960, whilst the second one ordered the creation of the Planning Department Unit. Moreover, seeking to ignite economic growth, Singapore looked for advice on industrial development, which materialized on a United Nations Development Programme (UNDP) mission in 1960. Headed by the Dutch industrialist Albert Winsemius and his secretary I.F. Tang, the team examined Singapore's industrial and manufacturing potential. One year later, in 1961, the Economic Development Board (EDB) was formed to solve the growing unemployment problem and attract foreign investment to help build the industrial base and promote economic development (Yap *et al.*, 2009).

In 1965, the 1955 Master Plan went through its first mandatory revision. Although approved with major changes in 1966, it was not extended beyond 1972 because of the assessment exercise carried out by the UN that would result in the Concept Plan 1971 (Yeung, 1973) and which responded much more to what was envisioned for the city-state.

Since attaining self-government, the government's decision to transform Singapore into a modern city was associated with a set of planning concepts. The notions of the 'Ring' and the 'Garden' city, two concepts from that time and still very much popular at present, gave shape to modern Singapore. Originally a Dutch urban development notion, the concept of the 'Ring City' was introduced to the island in 1963 by a UN planning team led by Otto Koenigsberger, a German architect-planner and housing advisor to the UN Economic Commission for Africa.

That very same year, an Urban Renewal Plan was prepared for a city that was already home to four million people. According to Waller (2001), the HDB's publication '50,000 Up' referred to this urban renewal process as a gradual demolition of virtually the old city's entire 1,500 acres to replace it with an integrated modern city. Planning for the 'Ring City', with subsequent revisions and modifications, radically altered the appearance of the city centre. It transformed it into a nucleus surrounded by a ring of new settlements connected with each other via a modern transportation system, and with the centre of the island being kept as a natural reserve and water catchment area. To this day, the primary and mature secondary forests at the central catchment and Bukit Timah nature reserves remain as Singapore's largest areas of tropical rainforest (see MND, 2002).

By 1965, Singapore had built 51,000 dwellings at a cost of M\$192 million,[2] each and every single one of them with water supply and sanitation facilities. The Queenstown area, the first of the new towns, had a population of 125,000 housed in 17,500 dwellings, in addition to shops, markets, schools, health clinics and community centres (Jensen, 1967).

Also in 1965, and as part of the political reality of the day in which Singapore was part of Malaysia, a new plan was developed favouring a regional approach to planning rather than as an isolated exercise. That plan had a regional element in it as it foresaw Singapore as a centre of manufacturing exports for Malaysia's industries and a main centre for industrial expansion. Thought to cover the region's development up until 1972, it planned for building housing infrastructure for an expected higher population growth, free migration between Singapore and Malaysia and a network of highways and a mass rapid transit system (MND, 1965). Given the uncertainty of those times and the changing socio-economic situation Singapore was going through, the 1965 Master Plan review was confined to the immediate 1963–1972 period.

With independence in 1965 and the separation from Malaysia, the period of political uncertainty ended, geographical boundaries were set within the main and offshore islands and the government went ahead with its pragmatic economic policies. As discussed in this chapter, domestic planning policies followed three lines of thought, which included pursuing the vision to build a global city and striving for national survival and achievement.

Focus on a 'clean and green' environment

The idea of making Singapore a clean and green 'Garden City' was first put forward by Lee Kuan Yew in 1963 (Waller, 2001). The term was borrowed from the planning concept coined by Ebenezer Howard in the early twentieth century in his 1901 *Garden Cities of Tomorrow* book (Guillot, 2008). In a Howardian sense, a typical 'Garden City' was to be a self-sufficient planned city with green belts for food production and physically disconnected from the industrial centre. The Singaporean Garden City concept does not quite follow the Howardian ideal of a low population density, self-sufficient settlement in which industry and farming provide a balanced environment nor does it incorporate Malthusian views and a rural-oriented idea (Choay and Merlin, 2005).

Quite differently, Singapore has developed its own style of a 'Garden City' with high-density development, high dependency emerging from its multiple economic connections as a global city and constantly aware of limited space and scarce natural resources. In fact, fully cognizant of its land, water and other natural resource constraints, the city-state has tried to work around and excel within these limitations. Therefore, in its development strategies, possible environmental damages caused in the process of economic growth have not been overlooked.

The conception of a 'Garden City' in 1963 called for another Urban Renewal Plan to be worked out. This new roadmap aimed at developing a modern city 'worthy of Singaporeans present and future as the "New York" of Malaysia' (HDB, 1965: 64). This spirit clearly indicated a kind of 'neo-utopian ideal', as Victor Savage (1991: 131–132) defines it. It is 'the Singaporean apotheosis for a live urban environment of anthropocentric human-nature close relationship'. In this context, the Land Acquisition Act of 1960 was highly significant as it empowered the PAP government to acquire land required for public housing. Without this law, the PAP government would have encountered the same obstacles as the British colonial government did in clearing squatters and pig farmers. It has been argued that legislations on the Woman's Charter, the liberalization of sterilization and abortion, labour laws and the 1960 Land Acquisition Act were all accepted by the citizens because they coincided with their interests and also because of the legitimacy enjoyed by the government itself (Leong, 1990).

Equally in 1963, when the urban planning concept was launched, the then Prime Minister Lee Kuan Yew started the 'tree planting campaign'. This was a far-sighted action that started sowing the seeds of the 'Garden City'. In fact, today's Singapore Green Plan traces its origin back to the tree planning campaign of the early 1960s (Koh, 1995). So much importance was given to this initiative that the 'Plant a Tree' campaign became an annual feature. As Yap *et al.* (2009) noted much later, Lee Kuan Yew and Goh Chok Tong were possibly the only prime ministers in the world who monitored the work of a National Parks Board Gardening Committee whose job was to plan, implement and tackle projects to improve the island. Singapore became, and remains, a 'Garden City' because the meticulous manicure of its trees, shrubs and flowers received the attention of the Prime Minister, no less.

Part of the island's improvement works, the 'clean and green' concept was essential for the water-scarce city-state. As Lee Kuan Yew recalls, '[o]ne compelling reason to have a clean Singapore is the need to collect as much as possible of our rainfall of 95 inches a year' (Lee, 2000: 205). The severe drought that affected the island also in 1963 was a wake-up call for the government and the newly formed Public

Utilities Board (PUB) to accelerate Singapore's plans to expand the catchment areas, reinforce existing reservoirs and make water resources free from pollution.

Soon after independence, a measure that significantly contributed to rapid urban renewal and housing development was the Land Requisition Act of 1966. With this Act, the government was not required to demonstrate public benefit for compulsory land acquisition by the state for any residential, commercial or industrial purpose, thus removing a major obstacle to redevelopment private actors and public authorities had previously faced. Later on, in 1969, the revision of the Act supported the recovery of rent-controlled premises for private redevelopment, an endeavour vigorously encouraged around the Singapore River by both these initiatives (Yeung, 1973).

Planning practices and expertise were brought from abroad with the State and City Planning Project (1967–1971) and institutionalized under the newly created State and City Planning Department. Most importantly, urban planning was not limited to spatial development and was fully integrated into a wider national social policy that allocated massive public investments in pursuing egalitarian goals (Teo, 1992). Part of the Ministry of National Development (MND), the Planning Department had the aim to bring about the optimization of available land resources and resolving conflicting development proposals in the overall interest of the State and for the common good (MND, 1985). The result was the birth of the Concept Plan of 1971.

Water management in the early years

When Singapore gained independence from Malaysia in 1965, celebrations were mute and attitudes were negative. Only two years before, the island had become part of the Malaya Federation. Its new 1965 Plan had charted the development path to be followed as part of greater political and geographical unit with a hinterland from where resources could be drawn and local products would have a market. Nevertheless, and even in those troubled days, a visibly distraught Lee Kuan Yew had the foresight to insist that the water agreement between Johor and Singapore be enshrined in the separation agreement between Singapore and Malaysia (Yap *et al.*, 2009).

The then Minister for Law and National Development, Edmund William Barker, drafted the agreement to separate Singapore from Malaysia, a document that was an amendment to the Malaysian Constitution allowing Singapore to separate and to proclaim its independence. Section 14 of the Constitution and the 1965 Malaysia (Singapore Amendment) Act constitutionally entrenched the much needed water supply source to Singapore. The insertion of the water clause in the Malaysian Constitution was a great achievement of the then Singapore leaders. This reads as follows:

> The Government of Singapore shall on and after Singapore Day abide by the terms and conditions of the Water Agreements dated Sept 1 1961 and Sept 29 1962, entered into between the City Council of Singapore and the Government of the State of Johore.

The Government of Malaysia shall guarantee that the Government of the State of Johore will on and after Singapore Day also abide by the terms and conditions of the said two Water Agreements.

(Malaysia, Act of Parliament, no. 53 of 1965. Constitution and Malaysia {Singapore Amendment} Act 1965. 9 August 1965)

The situation was precarious regarding inland water storage capacity, and recurrent droughts and floods inflicted hardship on the population and posed a major impediment for economic growth. In December 1969, floods had claimed five lives and caused damage worth $4.3 million (1969 prices). With 6,900 ha of flood-prone areas (about 12.75 per cent of the main island area at that time), the government started providing drainage in flood-prone areas as well as implementing flood prevention measures in low-lying areas to address the problem (Lim, 1997).

In the initial years after independence, financial limitations curbed planning and construction projects associated with water management. At that time, in response to the question of flood prevention, the then Minister for Law and National Development, E.W. Barker, mentioned finances as a limiting factor in constructing the drainage system and installing pumping stations (Hansard, 1969b). Nonetheless, despite the financial constraints, numerous drainage and flood alleviation schemes were subsequently approved such as the Bukit Timah Flood Alleviation Scheme, Phase 1; the Urban Renewable Drainage Scheme; the Sungei Kallang Improvement Scheme; the Kampong Kembangan and Siglap Improvement Scheme; and the concrete lining of Sungei Whampoa. In addition, actions like canal widening and the lining of Pelton Canal were completed. Budget constraints required engineers at the Public Works Department to come up with innovative and resourceful engineering methods. For example, the construction of a new canal for the 700 ha storm water diversion project from the Upper Bukit Timah catchment to Sungei Ulu Pandan was completed with a small budget of $7 million (Tan et al., 2009). By the early 1970s, rapid industrialization and economic growth took Singapore's per capita income second in Asia only after Japan. As the city-state became more affluent, it became much easier to plan and implement new flood alleviation measures and carry out water development projects.

In addition to the previous schemes, the government was also keen on developing an efficient sewage management system, which would improve people's quality of life and sustain the city-state's process of economic growth. As a result, a Sewerage Master Plan was conceived in the late 1960s, serving as a detailed guide for the development of used water facilities. The roadmap included macro-level projections of used water flows as well as micro-level design considerations of sewers, and layout of the sewerage facilities (Tan et al., 2009).

On 27 January 1970, E.W. Barker informed the House how sanitation facilities had expanded in Singapore, starting in 1949, when the population of 1.25 million was served mostly by modern sanitation. In 1969, 20 years later, with a population of two million, sanitation services were available to half of this number.

In a burgeoning economic climate and a time of rising societal aspirations and expectations, the Minister mentioned that he hoped the government would be able to extend this amenity to 95 per cent of its people within the next 15 years. In fact, Singapore achieved this goal in 1980, five years ahead of the foreseen time window (Turnbull, 2009).

Industrialization, environment and legislation

The Singapore government was highly conscious of the environmental pitfalls that accompanied industrialization. It thus developed its own integrated approach to environmental protection from a very early stage to ensure that industrial development did not take place at the expense of the natural environment by following an assemblage of strategies centred around prevention, enforcement and monitoring. With the government set to minimize environmental impact, managing competing land uses stemming from the need to develop industries, offices, housing, public space, forest reserves and water reservoirs was crucial for the island, which at that time had an area of only 580 km².

The city-state promptly adopted an integrated approach to land use planning. This strategy was carried out in consultation and coordination with all concerned government agencies, which shared not only the allocated resources but also took part in mitigating the adverse impacts of development on the environment, especially on water resources. This was a conscious effort on the part of the Singapore government and many early policy-makers that would receive tribute from the Minister for Environment Ahmad Mattar two decades later:

> In Singapore, we had recognized these problems early in our housing and industrialization programme. Industries are sited outside water catchments, and every new housing state is provided with central sewerage facilities conveying waste-water out of the water catchments. To do this is not without cost, but we have no other option. Our enforcement against illegal dumping of toxic waste has to be stringent. The penalty is a heavy fine for the first offence, and mandatory imprisonment for the repeat offender. Our present policies for environmental planning, and sewerage will continue.
>
> (Mattar, 1987: 19)

Singapore's early integrated land use management approach ensured all developments undertaken were located in designated areas to mitigate environmental impacts. It also made it a rule to incorporate pollution control measures in the designs of all new developments. Therefore, measures to control air, water and noise pollution, the management of hazardous substances, and the treatment and disposal of toxic waste were required to be examined and indicated clearly in the proposed plans of any development activity.

Relevant to this book's scope and from the very onset of its life as an independent nation, the island's water vulnerability challenged the government to adopt strategies well beyond water pollution control to encompass measures to maintain, reinforce, enhance and reuse its water resources. The pragmatic leadership understood the impacts land use, chemicals, toxins, air pollution and waste have on water and to avoid these activities and substances polluting the water sources, stringent laws were passed and strictly enforced.

In 1970, and to promote environmental responsibility and give full political backing to this initiative, Prime Minister Lee placed the first environmental agency, the Anti-Pollution Unit (APU), within his Office. This was as a response to the 'Cleary Report', prepared by Graham Cleary, a World Health Organization (WHO) consultant on air pollution who had recommended that the spirit of enforcement should be by persuasion and advice rather than coercion (Cleary, 1970).

Contrary to the previous suggestion, Lee Kuan Yew not only favoured stringent laws but also ensured their stricter enforcement, with 'command and control' tools being used to control pollution throughout the years. Despite expert advocacy against such measures as economically inefficient, Singapore has balanced environmental considerations and economic imperatives, avoiding the sort of environmental destruction many developing countries are experiencing as a result of policies aimed at attracting 'investment at all costs' (Hernandez and Johnston, 1993). Immediately after the Cleary Report and the formation of the Anti-Pollution Unit, The Clean Air Act (Chapter 45) 1971 was enacted. This piece of legislation gave the Director of the Air Pollution Control Unit more discretionary power to take measures against any air impurities that were being, or likely to be, emitted from any industrial or trade premises.

The government also enacted the 1968 Environmental Public Health Act, incorporating provisions from the 1963 Local Government Integration Ordinance on air pollution control and maintenance of public health. Under this Act, littering, spilling of noxious and offensive matter and spilling of earth in public were made criminal offences and strict penalties imposed on offenders. Enforcement procedures were designed to deal with any offence expeditiously and with minimal paperwork.

In 1968, before the Concept Plan was drafted in 1970, the notion of the 'Garden City' was officially presented during the introduction of the Environment Public Health Bill before the Parliament. At that time, the then Minister for Health, Chua Sian Chin, clearly stated how: 'The improvement in the quality of our urban environment and the transformation of Singapore into a garden city – a clean and green city – is the declared objective of the Government' (Guillot, 2008: 153).

In 1972, growing domestic and international focus on environmental matters led to the formation of the Ministry of the Environment (ENV) to specifically prevent all kinds of air and water pollution. The initiative to establish such a specialized ministry constituted a pioneering move in South-east Asia, and one that was further backed with newly formulated regulations.

In 1975, the Water Pollution Control and Drainage Act (Chapter 348) was enacted to control water pollution. Its overriding principle was that, wherever possible, effluents were to be discharged into sewers and their quality monitored and regulated. Part IV of this Act primarily addressed water pollution control for inland waters (rivers, streams, lakes or ponds) and made it a punishable offence to discharge 'any toxic substance into any inland water so as to be likely to give rise to an environmental hazard'. In addition, Section 4(1) of the 1976 Trade Effluent Regulations enabled the Director of the Water Pollution Control and Drainage to ensure that trade effluents were only discharged into sewers (Trade Effluent Regulations 1976, 1990). A detailed analysis of the regulatory instruments implemented in Singapore throughout the almost five decades since independence can be found in Chapter 3.

Finally, it is pertinent to briefly mention that relocating Singapore's population to HDB public apartments in the early 1960s rapidly cleared the island's central area and the Singapore River's surroundings. This measure chiefly brought about a considerable improvement to the population's living conditions. It is also important to stress how the redevelopment of the city and the necessary community resettlement in new HDB flats with proper waste disposal facilities reduced the amount of waste that used to be discharged on land and rivers. In short, the policy paved the way for modern urban development in Singapore.

Consolidation and first concept plan

In 1962, 1963 and 1965, UN consultant groups visited Singapore answering to a technical assistance request to deal with redevelopment problems in the city's central area. Overall, the different groups recommended replacing the previous 1958 Development Plan (Dix, 1959) with a more up-to-date document that could help lead a course of progressive development by organizing and directing the resources available to the government (Crooks, Michell, Peacock and Stewart Pty Ltd, 1971).

In 1971, a long-term, comprehensive Concept Plan was elaborated for Singapore's physical development for a population of four million,[3] mapping its vision for a period of 30 to 40 years. It articulated the strategic directions for land use and transportation and prescribed reviews every ten years. It took into account changing economic and population trends as well as land uses and transportation networks needed over the course of the city-state's physical growth (Crooks, Michell, Peacock and Stewart Pty Ltd, 1971). Accompanied by a Master Plan that detailed land use schemes, the Concept Plan entrenched the efficient and effective use of land resources so that the population's quality of life continued to improve even with, and due to, the continuous development of the city-state.

The 1971 Concept Plan provided a flexible framework within which detailed programmes could be prepared by the different government agencies. The document's main advantages were its flexibility and practicability since it could be adapted to future changing circumstances and implemented within known and predicted constraints imposed by governmental objectives, policies and resources, as well as market forces influencing the flow of private investment.

One of the fundamental aspects of this Concept Plan was the concept of the 'Ring' integrating approach, which as mentioned earlier in this chapter envisioned creating a development ring around the central catchment area. According to this notion, major industrial areas were to be located on the periphery surrounding corridors and major recreational areas were to be developed from the central catchment area through to the coast. The new towns were to be built around the central catchment area that covered around 46.6 km², where the MacRitchie, Peirce and Seletar protected catchments were located. This served the definite purpose of safeguarding water bodies from pollution while also easing population concentration from the central area. The protected catchments were left in their natural state as much as possible, not authorizing any development work to be carried out there. The Plan exercised careful control in the use of the area for recreational activities.

In essence, the Concept Plan was a guide for the development of the sort of infrastructure that would facilitate economic growth and would satisfy the housing requirements and basic social needs of the population. A new international airport at Changi, new towns, an expressway system, and a mass rapid transit system were among some of the projects to be undertaken. The Plan outlined how the central area was to be cleared of its residential population and industrial activities and be thoroughly renewed into an international financial, commercial and tourist centre (Teo, 1992) (see Figure 2.1). For this, high-density public housing was to cluster along the corridors and next to high capacity public transportation routes. This redevelopment was also essential to clean the waterways so vital for Singapore (Wong *et al.*, 2008).

While at the macro level, the Concept Plan guides Singapore's development through broad long-term strategies, at the meso level, the Master Plan – a medium term development plan – translates such goals into detailed activities. Both documents are the result the collaborative effort of various ministries and statutory boards and, through their competencies, support key action lines. As such, proper

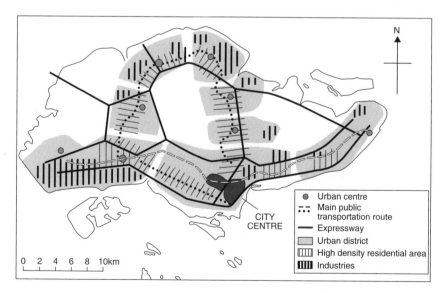

Figure 2.1 Concept Plan 1971 (Crooks, Michell, Peacock and Stewart Pty Ltd, 1971).

land use planning plays a major role in protecting water bodies and the environment. For example, stretches of land are set aside for sewerage, waste disposal and incineration facilities; similarly, polluting industries are grouped together and located away from residential areas. By clustering industries in one place, environmental protection measures can be properly taken into consideration and cost-effectively carried through by economies of scale (Seetoh and Ong, 2008).

Water supply strategies

Reflecting its strategic approach and comprehensive and wide encompassing scope, the Concept Plan 1971 was prompt in recognizing the urgency in addressing the issue of water scarcity, making supply the most critical among the public utility services (Crooks, Michell, Peacock and Stewart Pty Ltd, 1971). Moreover, the government was reluctant to continue depending so heavily on water imports and thus planned to develop additional catchments, a measure that was duly included in the Concept Plan itself. It was expected that, on completion, internal sources would meet 50 per cent of the total domestic water needs.

Also in 1971, a Water Planning Unit was set up under the Prime Minister's Office to assess the scope and feasibility to expand water supply. Based on the Unit's studies, which considered both conventional and unconventional water sources, the first Water Master Plan was drafted in 1972 and became a blueprint guiding Singapore's long-term development of its water resources. This document outlined a series of action lines to ensure diversified and adequate local water supplies and meet projected future demands, first aiming at creating urbanized catchments whilst following up on initiatives for water reuse and desalination (Tan *et al.*, 2009).

Planners soon grew aware of the fact that infrastructural developments and water imports were not enough to cover Singapore's freshwater demands. Supported by the highest political figure in the city-state, it became a national imperative to clean-up its highly polluted rivers and water bodies. The issue gained even more urgency as the Concept and Water Master Plans both stressed the need to develop unprotected catchments, especially in the urban areas.

As already stated elsewhere in this chapter and throughout the book, possible sources of water pollution had to be assertively closed. Animal husbandry activities near catchment areas were relocated, sanitation systems extended over the entire island and anti-pollution legislation introduced and effectively implemented. Tan *et al.* (2009) draw attention to how rapidly growing demand compelled the capture of every drop of water that was economically viable. The PUB then started its ambitious project of creating urbanized water catchments. A detailed account of the scheme followed to clean the Singapore River and the Kallang Basin can be found in Chapter 6.

Developing the Drainage Master Plan

The Concept Plan 1971 considered separate main drainage systems for each major catchment, facilitating the conservation of the runoff for water supply purposes.

When planning for these drainage and sewerage works, the maximum effluent reuse and pollution mitigation from industrial activities, farming and sand quarrying were identified as very important issues to be taken into consideration. The scheme also set freeing additional land as a crucial objective, and one that was to be attained by mitigating floods in at-risk areas that were thus unattractive for both development and investment. Likewise, treatment plants were foreseen for all main developed areas and the provision of suitable land and buffer zones around treatment plants envisaged (Crooks, Michell, Peacock and Stewart Pty Ltd, 1971).

In 1972, the Drainage Department was set up under the newly formed Ministry of the Environment (ENV). It was set the task of ensuring the construction of an effective land drainage system to protect national assets, improve public health and prevent and alleviate floods. The same year, a comprehensive Drainage Master Plan was drawn up and immediately implemented (Lim, 1997). For the following 25 years, the Drainage Department constructed some $2 billion worth of projects (PUB, personal communication, 2012).

Soon after the Plan was put into force, major flood alleviation projects were completed. At a cost of $37.4 million, the 2.5 km long project in the Stamford Canal was finalized in 1985 along with its 1.2 km tributary extending between Orange Grove Road and Cairnhill Road. The Bukit Timah flood alleviation scheme, mentioned earlier, was completed in 1972 to alleviate floods in the catchment's upper section. Flood alleviation works in the catchment's middle and downstream started in 1987 and concluded in 1990 after an investment of $175.8 million. With a disbursement of $22.2 million, the Siglap Canal and its tributary drains began operating in 1987. Finally, the drainage scheme at Tanjong Katong, which began in 1988, was inaugurated in 1993 with an investment of $33 million (Lim, 1997).

In view of rapid urbanization, many waterways were upgraded to facilitate increased storm water runoff. This work was done at the same time that HDB developed new towns or Jurong Town Corporation (JTC) developed industrial estates. Improvements were made in Sungei Bedok and Sungei Tampines to serve Bedok New Town and Tampines New Town; and also at Sungei Kallang to support developments in the new towns at Bishan, Ang Mo Kio and Toh Payoh. The Drainage Management Catchments can be seen in Figure 2.2.

The Drainage Department was also entrusted with the enforcement of the Water Pollution Control and Drainage Act (1975), the Surface Water Drainage Regulations (2007) and the Trade Effluent Regulations (1976). It thus had the authority to award the Certificate of Statutory Completion on development works carried out under approved building plans. This organization was likewise responsible for the monitoring of construction sites to ensure that construction works did not affect existing drains and that new ones were put in place according to cleared building plans.

The multiple drainage projects have helped reduce flood-prone areas by more than 95 per cent over the last few decades, even as urbanization has intensified over the same period of time. From the 3,178 ha of flood prone areas in 1970s, only 56 ha remained vulnerable in March 2011 (PUB, 2011b) (see Figure 2.3).

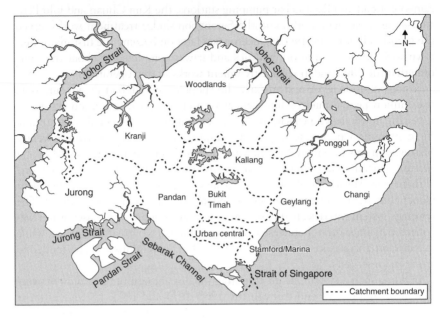

Figure 2.2 Demarcation of drainage management catchments (source: Lim, 1997: 190).

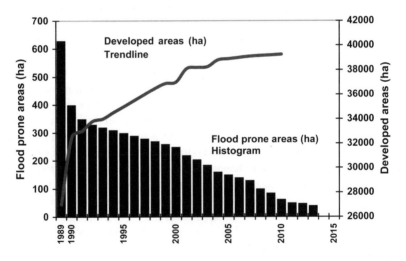

Figure 2.3 Reduction in flood-prone areas despite urbanization.

Developing the Used Water (or Wastewater) Master Plan

In the early years, the solution given to Singapore's sanitation problem was the nightsoil bucket collection service. This system remained in use until the late 1980s when it was finally replaced by alternative on-site sanitation systems. Colonial sanitation structures such as the Alexandra Sewage disposal works, the

Rangoon Road and Paya Lebar pumping stations, the Kim Chuan and Ulu Pandan sewage treatment works, and the Serangoon sludge treatment works served the island well until the 1960s. Queenstown later on became the first major new town with on-site sanitation facilities and it was completed at about the same time that the Ulu Pandan sewage treatment works started operating.

Rising pollution levels and an expanding population that had already surpassed two million people made a more effective and expanded sanitation system to cover the newly developed satellite towns an imperative, especially to minimize the contamination of waterways and coastal waters. Consequently, a Sewerage Master Plan was formulated in the early 1970s (later renamed the Used Water Master Plan), dividing Singapore into six used-water catchment zones, each one of them served by a centralized water reclamation plant (or wastewater treatment plant) compliant with international standards (Lawrence and Aziz, 1995). The sewerage system was designed based on a 'separate system' where used water was collected in a network of underground sewers that led to a treatment plant, while storm water and surface runoffs were collected in open drains and channelled to rivers and reservoirs.

Building such a system meant that a significant amount of financial resources were allocated to this end through the 1968–1973 and 1973–1978 investment programmes; expenses later recovered from consumers using cost-recovering mechanisms (Tan *et al.*, 2009). The sizeable investment and work put into building a robust sewerage infrastructure greatly contributed to the nearly universal coverage of modern sanitation services. Within 20 years, Singapore eradicated the use of bucket latrines and open defecation. With time, on-site sanitation became a service provided to the whole population.

Urban sprawl meant that following the old open trench method of sewer rehabilitation was not only going to be a time-consuming endeavour but one that would prove rather inconvenient to the public as well. Seeking to avoid these difficulties, and from the early 1980s, the PUB explored the use of 'trenchless' technologies, which saved enormous amounts of time and cost, and proved less disruptive for the public. Also at that time, when construction of the urban catchments was also beginning, leaks were identified as a major threat to the aesthetic and recreational quality of water bodies, leading the PUB to explore newer technologies and better sewer rehabilitation programmes (Tan *et al.*, 2009).

Urban development post-1980

By 1989, much of the infrastructural development considered under the 1971 Concept Plan and required for the increasingly prosperous and distinct urban city had already been developed. Waterways were clean and the 1985 Singapore River Development Plan was in progress. Most of the population was living in high-rise HDB flats, enjoying full access to piped water and modern sanitation. New urban estates, expressways and an efficient and expanding transportation system had been developed.

Moreover, the successful cleaning of urban water pollution sources made it possible to build urban water catchments in highly urbanized areas where residential, commercial and industrial complexes were to be found. In 1986, the first urban catchment at Sungei Seletar-Bedok, a technically complex storm water collection system, was developed. As part of this pioneer urban catchment, Sungei Seletar was dammed to form the Lower Seletar Reservoir and raw water piped to the Bedok Reservoir. This was constructed out of a sand quarry, and which receives whatever is caught at storm water abstraction ponds in highly urbanized towns, namely Bedok, Yan Kit and Tampines (Tan, 2009).

It is important to stress the decisive role integrated planning and agency coordination played in all of these efforts. For instance, in the development of the Bedok catchment, Urban Redevelopment Authority's (URA) far-sighted planning kept polluting developments away from water sources ten to 15 years prior to the construction of any reservoir scheme. This area was reserved for light industry and residential purposes only and it has an extensive sewerage network jointly planned by the PUB, URA and HDB (Tan *et al.*, 2009).

On a concretely more social level, achievements in health and living conditions were remarkable. For example, the steady decline in indigenous typhoid fever recorded from 1980 (142 deaths) to 1989 (33 deaths) was mainly due to improved food hygiene standards, the better control of public food handlers and the more extensive supply of clean water and widespread access to modern sanitation services. The proportion of the population with such amenities went from 76 per cent in 1975 to almost 100 per cent in 1989, an outstanding feat given how fast the population was growing. In addition, street hawkers were relocated to markets and food centres where modern water supply and sanitation facilities were available. By 1989, only 174 (0.8 per cent) out of 21,097 hawkers operated outdoors compared to the 22,574 street vendors that existed in 1975 (Yew *et al.*, 1993). With these achievements under its belt, the city-state then looked ahead to making qualitative improvements to its public spaces and endowing them with a distinct identity.

In guiding this physical and aesthetic improvement, safeguarding environmental norms, and following through the vision of a 'Garden City', the HDB collaborated with the Parks and Recreation Department to provide landscaped open-space greenery by planting and maintaining tropical trees. By then, implementing the State and City Planning Project and the renewal of the Central Area posed less of a challenge, greatly assisted by the fact that water was being properly managed. Land acquisition and a wide range of infrastructural development such as roads, sewerage, drainage, telecommunications, etc., helped transform the once rundown area into an attractive and vibrant commercial and business district.

Water and development in first world Singapore

In a few decades, the city-state built a well-deserved sense of national pride and achievement based on the many achievements it had made in the pursuit of the development path it embarked on since independence. This sentiment received a prestigious boost when, in 1985, the International Monetary Fund declared Singapore a first world country. According to the Ministry of Trade and Industry (MTI), the official foreign reserves stood at $33.2 billion in 1988 (Quah, 1991). That August, the government decided to call for a general election 15 months ahead of time and the PAP campaigned under the slogan of 'More Good Years'. Singapore was ready for the next step towards urban excellence. This was the last election for which most of the first generation of political leaders ran before retiring from the political scene only after passing an efficient and effective administration and an affluent economy to the next generation.

In 1985, Singapore's robust economy experienced a slight downturn but soon recovered, recording high annual growth rates from 1987 onwards. Its diversified growth base meant it was hardly affected by the first Gulf War, which conversely drove the country's main trading partner, the United States, into recession. Before the end of the 1980s, the city-state consistently had budget surpluses, accumulated substantial foreign exchange reserves and was ready to take its economy in a new direction (Turnbull, 2009).

Striving to always look forward, the country's consciousness of quality and quest for identity pervaded all ways of life, including water, the economy and even public housing (Cheong-Chua, 1995). In the 1960s, the country had to address two priority matters. It had to assure water supply, an issue that grew even more crucial after independence in 1965, and, in economic terms, its first priority was to attract investment and generate employment to pursue the advised industrialization model. In the 1970s, the water focus entirely shifted and moved towards self-sufficiency by exploring conventional and non-conventional means. In this quest, Singapore tried technologies that were not economically viable at that time such as desalination and recycling of used water and that only with time matured as feasible solutions. At that point in time, the economy started awarding an ever larger and more relevant importance to diversifying its income base and upgrading its workers' skills.

This paradigm evolution continued and with it the responses given to pervasive and new challenges. By the 1980s, the country had put in place the infrastructure required to efficiently deliver clean water; it then immersed itself in the cleaning of waterways, which constituted a big improvement towards making the inland water resources pollution free; and it could claim that severe floods had become a thing of the past. The economic prosperity that developed over the years established and firmly rooted Singapore as a first world country. Later that decade, the city-state embarked on a 'second industrial revolution' involving the use of increasingly sophisticated and cutting-edge technology and setting the foundations to build a knowledge-based economy (Turnbull, 2009).

Unlike many big metropolises worldwide, Singapore's industrialization process has not led to environmental decay. Defying the trend, economic development and industrialization have not been inversely related to environmental quality. Quite the contrary, the city-state has demonstrated that if designed, built and operated in environmentally sustainable ways, the vast ability cities have to generate immense finance capital could be harnessed to protect and improve their urban landscape. Indeed, part of Singapore's environmental profile can be attributed to changes in its economic structure. In 1961, the natural resource intensive and heavily polluting food, printing, publishing and wood manufacturing sectors accounted for 40 per cent of total industrial employment. Thirty years later, these activities employed only 8 per cent of industrial labour, while the share drawn by electronics and electrical appliances had risen to 40 per cent (Leitmann, 2000 in Ooi, 2005: 81).

There is no doubt that pragmatic policies and attractive schemes to bring in foreign companies that could rely on quality and efficient infrastructure led to the great economic development witnessed by the city-state. This material prosperity stood as the backbone of Singapore's nation-building programme with the PUB playing an underlying role in guaranteeing that self-sufficiency would be attained and services were provided expediently and efficiently as part of this effort to promote economic growth. Just to exemplify the extent to which the PUB exercised its many tasks, and clearly appropriate given this book's central theme, in the 20 years spanning 1972 to 1992, the sewerage network first increased from 810 km (in 1972) to 1,600 km (in 1982) to then reach 2,340 km (in 1992). The number of pumping stations saw a parallel surge, going from 46 premises (in 1972) to 128 (in 1982) and totalling 136 at the end of this timeframe (Lawrence and Aziz, 1995).

With an affluent society, high standards of living, total urbanization and intensified industrialization, the demand for water grew rapidly over the years. Between 1963 and 1993, the population rose from 1.8 to 3.3 million driving the steady expansion in both the domestic and industrial demand for water. Between 1965 and 1993, consumption of water more than doubled from 75 l/capita/day to 173 l/capita/day. On their part, the industrial and commercial sectors had prospered to the extent that, in 1993, their share of water consumption had reached 36.25 per cent of total volume sold, a 144 per cent increase from the 1983 level, and 264 per cent from what was used in 1973. To further illustrate the magnitude to which water demand was tied and driven by economic imperatives, regardless of how advanced the established industries were, the post-1990 electronic revolution put an unprecedented toll on water supply. For example, the fabrication of silicon wafers used in the semi-conductor industry was a water-intensive and highly specialized operation. One plant alone operated with 600 gallons of water per minute, or what comes to account for 0.39 per cent of total water consumption in Singapore (Ooi, n.d.).

Naturally, with economic success, the country's built landscape was substantially and expeditiously transformed. Intensive use of limited land resources and high opportunity costs significantly reduced the number and size of farms, forests and marshes, while the space occupied by constructed areas increased from 18.5 per cent to 48.6 per cent between 1950 and 1993 (Low, 1997). In such a scenario, it was imperative to find different ways of making the city as green as possible and with water to be seen everywhere.

After the consolidation of the economy and the completion of the first cycle of urban development, policy-makers began considering the second round of urban improvement aiming at enhancing quality, providing for variety, achieving excellence in architecture and innovating in the design of urban spaces. Overall, it was also sought to make a more efficient use of the country's natural assets, namely its water bodies, tropical weather and abundant vegetation. For an island, finding ways of making water a part of urban development was perceived as imperative. The first steps would then be to seize the many advantages presented by the waterfronts of Marina South, the Singapore River, the Kallang Basin, Tanjong Rhu and Kampong Bugis.

Shortly after the river clean-up exercise in 1987, visionary Prime Minister Lee Kuan Yew steered the way in this direction during a television interview:

> In twenty years, it is possible that there could be breakthroughs in technologies, both anti-pollution and filtration, and then we dam up or put a barrage at the mouth of the Marina – the neck that joins the sea – and we will have a huge freshwater lake. The advantages are obvious. One: A large strategic reserve of water – fresh water – for use in emergency: a drought, or some such period. Second, it will help flood control because at high tides – exceptional high tides – which happen about two periods a year, if they coincide with heavy rain, the three rivers and canals will flood parts of the city. Now with the barrage, we can control the flooding. And with the barrage, the water level can be held steady. We need never [sic] have low tides. So the recreational use and scenic effect would be greatly improved. And it is possible in another twenty years, and therefore, we should keep on improving the quality of water.
>
> (Hon, 1990: 194)

The 1991 Concept Plan

The infrastructure the country needed and that was spelled out in the Concept Plan of 1971 was all developed already by 1989. That same year, to start mapping Singapore's future urban vision, the URA merged with the Planning Department and Research & Strategic Unit of MND. The new body became the national planning and conservation authority and allocated many more resources to lead the island's physical development into the year 2000 and beyond.[4]

In 1989, Liu Thai Ker, then new URA Chief Executive, was asked to review the 1971 Concept Plan. This resulted in new policies and direction for the future environment of the city-state. It considered that many social and economic changes had occurred since the previous Concept Plan had been developed: population had considerably increased, industrial developments posed new demands and extra land was required to make possible a better lifestyle for more recreation-conscious citizens (Waller, 2001). Already between 1979 and 1989, national expenditure on recreational activities and education had increased three-fold but with the turn of the millennium, Singapore needed new development strategies to meet rising expectations (URA, 1999).

Planners felt they had to bring housing closer to the water and facilitate commercial and other leisure activities at waterfront locations: the sea, at rivers and even canals. The work carried out under previous plans had reasonably landscaped and cleaned the rivers and had made them safe for water sports. New beaches near the downtown waterfront would now provide attractive venues for boating and water-skiing. Thus, a combination of vision and planning skills initiated the task of transforming Singapore into a vast tropical resort and named it 'A Tropical City of Excellence'. Consequently, the Concept Plan 1991 included three staging plans for the years 2000, 2010 and the unknown year 'X' when population would reach four million (URA, 1991). These staging plans would then be used as a reference point to help in the preparation of the 55 detailed Development Guide Plans (DGPs) that would map out concise initiatives for the entire city-state. These 55 DGPs would gradually replace the 1985 Master Plan.

More detailed and flexible than its 1971 predecessor, the 1991 Concept Plan retained the idea of laying out a framework for the island's physical development according to Singapore's vision of and for its future (Waller, 2001). By that time, Singapore had already become the first developed country in the equatorial belt, so it initiated the task of creating an international investment hub and leading it towards becoming 'A Tropical City of Excellence' (Liu, 1997). This planning roadmap analyzed in detail a variety of issues like economic infrastructure, transportation networks, housing, green networks and waterways, social and cultural facilities, and pollution control measures.

For the year 2000, the goal was to consolidate commercial and housing developments along existing mass rapid transit (MRT) routes so that the growing population had a transport system to support these new activities. It also planned works for massive new regional centres at Tampines, Seletar, Woodlands and Jurong East; as well as new industrial areas and business parks. By 2010, the north-east corridor was to be expanded, opening new residential neighbourhoods and a new light rail transit scheme (LRT). In the central area, Marina South was to be developed as part of the new downtown district that would allow for more user-friendly areas, waterside promenades, parks, hotels, shops and some residences.

For the year 'X', the focus would shift offshore to the east and south-east and would rely heavily on reclamation. It listed building housing and recreational facilities running from Marina East to Changi in 'Long island', after the area was reclaimed from the east coast. The Plan also foresaw fully constructed housing at Marina South, Marina East and the downtown core served by the extension of the MRT lines. It was also planned to radically change the landscape in the currently undeveloped islands of Pulau Ubin and Pulau Tekong, using them for new town developments connected to the MRT system. Including the aforementioned projects in the two islands, a total of ten new towns were planned. The Plan also considered opening more resorts, beaches and marinas.

Overall, the Concept Plan 1991 was intended to change the fabric of the society 'slowly, subtly but surely to a richer colour' (URA, 1991: 16). It thus considered the improvement of existing waterways such as reservoirs, rivers and canals into more aesthetically pleasing waterfront environments for recreational conviviality. In addition, selected coastal areas would be opened up for marinas and the 2,000 ha existing nature reserves in the central catchment area would be preserved. Apart from such green pockets, other select areas of ecological merit would be conserved. The central area was to be transformed into both an international business hub and a new downtown core displaying the most modern hotels, offices, shops and nightlife in the waters around Marina Bay (Cheong-Chua, 1995).

The Concept Plan 1991 emphatically established how judicious planning would turn water vulnerability into a strength: the scarce resource was to be used as a tool to create the new vision of Singapore and its use and impact was to be universal in the development of new areas either in the main or the underdeveloped islands. The URA vision document puts a great deal of different but bold emphasis on water where Singapore was envisaged as:

> A city surrounded by regional centers, each serving up to 800,000 people and each allowing Singaporeans to have jobs closer to home. An island with an increased sense of 'island-ness' – more beaches, marinas, resorts and possibly entertainment parks as well as better access to an attractive coastline and a city that embraces the waterline more closely as a signal of its island heritage. Singapore will be cloaked in greenery, both manicured by man and protected tracts of natural growth and with water bodies woven into the landscape.
>
> (URA, 1991: 4)

The importance of water in enhancing the value of real estate and creating business opportunities was clear for everyone to see. New market opportunities and prospects were created for the private sector to invest and transform dilapidated areas or virgin land into highly valuable assets. The government supported private initiatives with timely land releases and infrastructural development. Areas developed next to the Singapore River, Tanjong Rhu and Marina Bay areas are examples of this public–private collaboration (Cheong, 2008). Quickly, the URA and PUB turned to working closely together to utilize canals and inland reservoirs to create attractive and landscaped 'lakes and streams'. As a result, more leisure space was made available and real estate values soared.

Beyond the upgrading and redevelopment of individual housing areas, HDB also drew up a blueprint to renew and remake the HDB heartland. Under this roadmap, Singapore's public housing estates were to be totally transformed over the next 20 to 30 years. Waterfront housing was to be built in Punggol, middle-aged estates developed in Yishun, and new innovative design concepts like 'Housing in a Park' and Sky Gardens adopted. It should thus come as no surprise that the new business parks were planned to ensure occupants had an integrated work, home and recreation environment. These 'science habitats' were to offer high-quality housing and recreational facilities to attract and retain the talent (URA, 1991: 21). Singaporeans could look forward to living in even more lively and sustainable communities in the coming years (Tay, 2008).

Water management from 1991 onwards

Urban development and its emphasis on recreation and better quality of life could give the impression that there has been a shift from immediate water issues to more aesthetic ones. Nevertheless, in practice, core water concerns have never ceased to be a priority. Water has consistently been recognized as a strategic resource and its conservation has always been considered as a vital national security aspect (Hansard, 1989b). Quite on the contrary, the PUB has continued working towards an ever more efficient provision of water services and better planned and managed water resources, whilst trying to keep a balance between the lifestyle that people would expect to lead in a developed country and the need for water conservation.

Concurrently with the rapid development of the city-state, appropriate control strategies were adopted, older legislation and regulations were amended and new ones were drafted. Even though by 1980 a basic legal framework was already in place to meet Singapore's environmental needs, new issues coming up as a result of development activities demanded more suitable laws. Some examples include the Water Pollution Control and Drainage Act (Chapter 348) 1975 which was repealed and relevant powers were streamlined into the Sewerage and Drainage Act (SDA, now under Chapter 294) administered and enforced by the PUB and the Environmental Pollution Control Act (now known as the Environmental Protection and Management Act (EPMA), Chapter 94A) as well as their regulations, which are administered and enforced by the PUB.[5] For a detailed analysis on legislation, see Chapter 3.

Throughout the years, strong emphasis has also been put on water conservation campaigns. Nonetheless, after 25 years of carrying out awareness raising efforts, it was clear that such initiatives were not enough to conserve water. It was then decided to make use of every possible source and technological device and process to achieve this goal. In 1997, water pricing was revised not only to cover the full cost of production and supply, but also to reflect the higher cost of alternative sources. This sent a strong signal to the population and it encouraged the adoption of technical solutions for water fittings and their economic use (Tan *et al.*, 2009). All non-domestic premises were required to install water-saving devices such as self-closing delayed action taps and constant flow regulators. Since 1992, low capacity flushing cisterns employing 3.5 to 4.5 litres per discharge were installed in all new public housing apartments, a mandatory measure for all new premises since April 1997. A specific account of water demand management is presented in Chapter 4.

In 1983, and for ten years, the PUB started a $55 million pipe replacement programme. It involved changing all unlined water mains with cement-lined ductile iron and galvanized iron pipes with corrosion-resistant stainless steel and copper pipes. Some 182 kilometres of old unlined cast iron water mains and 75,000 of old galvanized iron connecting pipes were removed. This project ensured that water supply remained at its best and that leakage losses were kept to a minimum (PUB, 1992). This also made it possible for the PUB to have one of the lowest ratios of unaccounted-for water in the world, averaging 5 per cent in 2011 compared to 10.6 per cent in 1989 (see Figure 2.4).

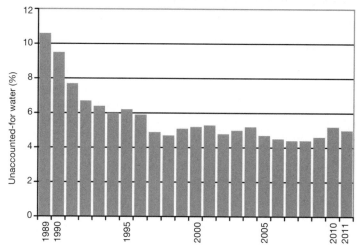

Figure 2.4 Unaccounted-for water (source: Public Utilities Board, 2011, personal communication).

Competing land uses in a fast-growing Singapore made land a premium commodity and thus its efficient use became a priority. Innovations and technological advancement were vigorously pursued with this aim in mind. For example, conventional water reclamation plants with open tanks had a 1 km-long buffer zone where very limited development was allowed. In the 1990s, the majority of them were covered, odour treatment facilities added and the buffer zone reduced to 500 m. Later on, all water reclamation plans were covered and made even more compact.

Urban development and the aesthetic approach of water resources

The use of water bodies aligned with the aims stated in the Concept Plan 1991 to make Singapore a more aesthetically pleasing place that offers a better quality of life (URA, 1993). Thoroughly recognizing the work undertaken with regards to water bodies that took place in the first two phases of Singapore's growth, the then Minister for National Development, Richard Hu, stressed how:

> Water bodies are ubiquitous feature of the Singapore landscape. Whether canals, drains, reservoirs or stormwater collection ponds, their development has allowed us (Singapore) to overcome the physical impediments of high rainfall and low topography to develop a modern and prosperous city state.
> (URA, 1993: 1)

In 1988, the First Deputy Prime Minister (who would later become Prime Minister), Goh Chok Tong, issued a discussion green paper, 'Action for Change', which after long deliberations was named as 'The Next Lap'. This document was the culmination of the aspirations of innovative and hard-working Singapore-ans to make the city-state 'more prosperous, gracious and interesting' (Turnbull, 2009: 350).

Opulence, sophistication, widespread well-being and thorough environment awareness made the development of attractive water bodies an essential part of shaping Singapore as a tropical city of excellence. Over the years, subtly and surely, urban planners, designers and the public alike started to see water bod-ies as assets rather than just storm water drains and collection points. Prosper-ity increased the demand for attractive and recreational life-quality enhancing water bodies and a series of measures followed in the pursuit of this vision. The Water Bodies Design Panel was formed in November 1989 and was comprised of representatives from both the public and private sector. Its main role was to look into the aesthetic improvement of water bodies when planning for further devel-opment. The importance attached to water sources can be gauged from the fact that the term 'water bodies' includes not only major watercourses like rivers and canals, but also minor watercourses like streams and drains as well as all kinds of reservoirs, storm water collection ponds, fountains, waterfalls and water frontages suitable for recreational purposes (URA, 1993).

After cleaning the Singapore River, a comprehensive Plan was developed by the URA and the Singapore Tourism Board (STB) in coordination with other departments and statutory bodies. The URA took care of architectural design, including the conservation and preservation of patrimony shop-houses and other buildings of heritage value as part of the vision to create a river of 'excitement' that blended elements of urban history with contemporary land uses. The STB followed with various activities and the use of public spaces to attract tourists. Singapore River was chosen among one of the 11 thematic zones identified in the Tourism Master Plan (Tourism 21) seeking to project Singapore as a tourism capital in the twenty-first century (STB, 1996).

While Singapore River was developed as a 'River of History and Entertain-ment', the URA also drafted the Master Plan for the Urban Waterfronts at Marina Bay and Kallang Basin to guide and integrate water-based and waterfront developments in those two areas. Together with the 1985 Singapore River Plan, this scheme charted out a comprehensive strategy for the island's three urban water assets covering some 265 ha. The Plan also provided the basis for the imple-mentation programme, eliminating incompatible uses and identifying proposals to be carried out in the near future (URA, 1989).

In the quest for making a more efficient and effective use of water bodies as a feature in urban design, 'urban integration' became a key notion. It was also used to describe the urban design concept for the area around the Kallang Basin, an important water body surrounded by Crawford, Kampang Bugis and Stadium (URA, 1997). The developments taking place during 'The Next Lap' have firmly established water as an integral part of Singaporean urbanity.

Further illustrating agency collaborative endeavours, the 1992 Singapore Green Plan, prepared by the ENV, came to complement the aforementioned aesthetic improvements. The Green Plan charted the strategic directions to be adopted to achieve its goal of sustainable development and was presented at the June 1992 Rio Earth Summit in Brazil (Ministry of the Environment, 1993). To keep the Plan relevant amidst the changing economic and environmental landscapes, it was reviewed and a revised 2012 Plan launched in August 2002 (Lee, 2008).

Looking ahead, the Plan saw Singapore's evolution into a 'City in a Garden', a bustling metropolis in tropical greenery. It is a challenging goal to green a small city-state where space is understandably limited in a land area of little more than 700 km^2 and a population of more than five million people. Nonetheless, not all green areas need to be compromised by economic and population growth. Even when land is scarce, Singapore has been able to commit 9 per cent of the total land area to parks and nature reserves by carefully planning for it. Between 1986 and 2007, despite population surging by 68 per cent from 2.7 to 4.6 million, the country's green cover grew from 35.7 per cent to 46.5 per cent (Ng, 2008).

Adding blue to green

The importance of greenery for a quality living environment has been stressed in Singapore's 2003 Master Plan. Accordingly, efforts are being made to bring people closer to nature and, where possible, to integrate nature areas within parks, also planning for an island-wide network of green links to connect parks and water bodies with residential areas (PUB, 2006).

In line with Prime Minister Lee Hsien Loong's vision of Singapore as a city where greenery and water would thrive, waterways would be integrated with parks to create new community spaces and the waterfront brought to the heartlands. To make these plans a reality, the PUB developed its Active, Beautiful and Clean (ABC) Waters programme with 20 projects already implemented. An example is the creation of a waterfront town at Punggol by damming up the mouths of Sungei Punggol and Sungei Serangoon to form two freshwater lakes and then joining them through a man-made river running through the estate and town centre, thus creating opportunities for waterside activities. This ambitious scheme aims at transforming Singapore into a 'City of Gardens and Water' as envisioned in the 2001 Concept Plan review (PUB, 2006).

The 2012 Singapore Green Plan, launched ten years earlier, charts the country's pathway to achieve long-term environmental sustainability within a decade, also setting out the broad directions and the strategic thrusts that will help ensure this. It focuses on six areas: clean air and climate change; water; waste management; public health; conserving nature; and international environmental relations. It was reviewed in 2005 and an updated edition was published in February 2006 (Lee, 2008).[6]

The importance of planning for the long term has also been understood and reflected in the blueprint presented by the Inter-Ministerial Committee on Sustainable Development (IMCSD), which has established key goals for 2030 to guide Singapore towards a more lively and liveable city. This 20-year timeframe vision has also set intermediate goals to be reviewed in 2020 to assess the progress achieved and correct the course if necessary. As part of this scheme, the role of water has been expressed as an aim towards self-sufficiency and greater efficiency. The targets seek to reduce total domestic water consumption from 153 l/capita/day in 2011 to 147 l/capita/day by 2020 and even to lower levels up to 140 l/capita/ day by 2030. Additionally, it looks into opening 820 ha of reservoirs and 90 km of waterways for recreational activities by 2020, reaching 900 ha of reservoirs and 100 km of recreational waterways by 2030 (MEWR and MND, 2009).

Amidst growing prosperity, Singaporean leaders are apprehensive about the devaluation of a scarce resource like water. In its attempt to raise awareness among people about the value of scarce water resources, the PUB has conducted many education and outreach programmes. These efforts are comprehensively looked into in Chapter 5.

Final thoughts

The issues presented in this chapter show not only the remarkable maturity of the country's political leadership but also flexibility and open-mindedness in dealing with new times and new challenges. A most important factor in the success of the city-state has been the pragmatic approach these leaders have followed in developing public policies, including those extended to water management and urban development. Even when decision-making has been heavily centralized, measures have remained matter-of-fact, sensible and looking out for lasting benefits as circumstances and needs have changed. Policy-makers have not been wedded to any paradigm, waging instead to achieve maximum long-term benefits for the city-state and its people.

The paramount importance of water for the city-state's survival has been reflected in the numerous decisions the political leadership has taken throughout the years. The dynamic framework for decision-making has been best captured and distilled in the different concept and master plans where goals looking to the future have been set by a visionary leadership who have placed the overall well-being of the nation as the matter taking precedence over all others. A nation's priorities and goals change with ever transforming social-political and economic conditions and the island has been able to identify, respond and also anticipate many of the changes and challenges brought about by time and development. Singapore's experiences show that it is the situation or the problem that defines the solution and not vice versa.

Forward-looking views, holistic approach, flexible planning, innovative management and search for technological advances have been a hallmark of Singapore's leadership, with the water story set in the context of urban development being a clear example of it. On the one hand, political sharpness has made it possible to ensure regular supply from foreign sources; on the other hand, it has been the government's endeavour to use the best of practices, instruments and technology to develop water sources and make them available to their domestic, commercial and industrial base as well as to the environment.

In its early development phase, Singapore tried to move towards self-sufficiency in water by enhancing its indigenous capacity and by cleaning its water bodies. The construction and expansion of catchment areas and reservoirs shaped urban development and facilitated maximum conservation and regulation of storm and inland water. At present, the policy is to bring water closer to its population in an attempt to conserve the scarce resource and promote social cohesiveness and sense of belonging. This is almost akin to the spiritual significance awarded to water in many cultures around the world and the restoration of water as symbol of wealth in Chinese philosophy.

Finally, looking at the country's journey towards sustainable development through land-use and water lenses, one can conclude that the holistic management of land and water resources has placed the city-state on track for security, self-sufficiency and sustainability. These are commendable objectives that will only rise in importance in the future.

3 Regulatory instruments and institutions for water pollution control

Introduction

Since independence in 1965, the government declared the objective of transforming Singapore into one of the cleanest cities in the world, a national priority second only to defence and economic development (Lee, 2011). It was considered that a clean city-state would not only be of great economic advantage that would attract foreign investment but would also bring in tourism. It would also be a great catalyst in raising the population's morale and civic pride so necessary to achieve higher performance standards (Chua, 1973).

From that first year as an independent nation, Singapore developed legal and regulatory frameworks to control increasing air, water and land pollution. Almost 50 years later, the success of these schemes is evident as the city-state has an environment that compares favourably among those of the best cities of the world.[1] The success is even more noteworthy because Singapore has been able to attract significant amounts of foreign investment in spite of its strict environmental regulations.

The island's early environmental policies back in the 1970s focused on air and water pollution, mirroring similar efforts all around the world. Following the Clean Air Act 1970 and Clean Water Act 1972 adopted by the United States and the measures passed by Japan's 'Pollution Diet' in 1970 (OECD, 2008), Singapore enacted its Clean Air Act in 1971, and the Water Pollution Control and Drainage Act (WPCDA) in 1975.

From the nation's onset, the government decided to incorporate environmental considerations in its development plans to minimize pollution and mitigate the overall impact on the environment (PCD, 2007). Throughout the years, institutions like the Public Utilities Board (PUB), the Anti-Pollution Unit (APU), the Economic Development Board (EDB), Jurong Town Council (JTC) and the Ministry of the Environment (ENV) have played a very important role in drawing in foreign investment and building up industrial infrastructure. These processes have been undertaken at the same time that environmental considerations have become an integral part in the overall course of industrial development.

From its inception, the ENV (now Ministry of Environment and Water Resources, MEWR) has been an essential and equal partner with the EDB and JTC on industrial development policy and decision-making. In the case of the

APU and ENV's Pollution Control Department, immediately after their creation in 1970 and 1972 respectively, their rigorous inspection and enforcement programmes made them competent, command-and-control regulatory agencies. Strict legislations and regulations were then passed to regulate pollution, requiring industries to obtain written permission before operating or acquiring facilities, which could potentially pollute the environment.

Regarding water pollution control, Singapore is very different from many other countries in at least three main respects. First, trade effluent discharge standards adhere to strict norms that do not allow watercourses to remain 'moderately polluted' as many other countries do. Second, the city-state has kept its laws relevant to the changing needs by regularly amending them and enacting new ones. Third, it has maintained the stringent enforcement of its laws, using vigilant monitoring as an important tool.

This chapter analyzes the legal and regulatory instruments for water pollution control that have been developed in Singapore over the years, as well as the role government institutions have played in their implementation. We conclude that the realization that the environment is a strategic element for security has made the city-state consider it as an integral part of its development policies.

Roles of institutions in policy implementation

At the time of independence, Singapore's economic and environmental future was uncertain given the declining export trade in staples, growing unemployment, limited manufacturing capacity, the collapse of the Malay Federation and the planned withdrawal of the British military. The environment was characterized by smoky and dirty air from wood-processing industries, open burning, and lax emission standards; and water was highly polluted with sewage, textile dyeing, as well as with pig, duck and all sorts of domestic and industrial waste. In the initial years, political turmoil also made it difficult to pursue long-term environmental strategies but as soon as some level of stability was attained, the government sought to make significant environmental improvements.

From very early on, the Singapore government adopted an environmental thinking that stressed how environmental problems could be prevented if appropriate controls were imposed. Therefore, from the very beginning, whenever new infrastructure has been planned, a holistic approach has always prevailed with environmental considerations being incorporated at land use planning level to minimize pollution and mitigate the project's impact on the nearby environment (PCD, 2007). In addition to proper land use planning, some additional fundamental criteria have also been applied to approve or disapprove of a proposed industrial activity: water usage; type of waste to be produced, and thus possible impact on the environment; whether the industry is of high value or not; and how skill intensive it is.

Institutions like the PUB, APU, EDB, JTC and ENV have played important facilitating roles in attracting foreign investors and building infrastructure, as well as in integrating environmental considerations in industrial developments.

For example, from its inception, the ENV has been an essential and equal partner to the EDB and JTC in deciding industrial development policies and their implementation. If the EDB finds that the objectives of upcoming industries are consistent with Singapore's economic goals, it goes ahead with pragmatic promotional activities for such endeavours. The JTC then has to provide factory space and infrastructural services. The ENV's role is to check that the new industrial ventures comply with the environmental norms, assessing in detail production processes and raw materials used as well as waste and emissions generated. In fact, it is only after ENV assessment and clearance that the EDB can offer promotional privileges to the industries. If the ENV does not approve of the proposed investment or abatement plans and if it disagrees with the EDB, the final resolution is taken by the cabinet.

Even after the ENV approves of a project, the interagency coordination machinery continues working to locate the specific industry in the most appropriate place. Once the ENV and EDB have given their approval, the JTC has to find the most suitable place for the industrial complex taking into consideration Singapore's Land Use Master Plan, which has its own environmental requirements.[2] The ENV classifies industrial activities according to their environmental impact, a parameter used to locate the industries and fix buffer zones.

For those industries that decide to construct their own facilities, the building plan needs to be approved by the Building Control Division (BCD), part of the Public Works Department and the ENV's Central Building Plan Unit (CBPU). The ENV further ensures that as part of the completed premises, environmental facilities have been installed and are operated properly. Only then can the BCD issue a temporary occupancy permit for the industry to begin its operations. The objective is that the entire process, which requires constant discussions and coordination among the various agencies involved, ensures industrialization without deterioration of the environment. For example, from 2007 to 2009, the CBPU processed an average of 9,596 yearly plans for residential and industrial developments as well as 3,543 applications for industries allocation in the JTC, HDB or private industrial estates (PCD, 2007, 2008, 2009).

Immediately after their inception in 1970 and 1972 respectively, the APU and the Pollution Control Department became competent command-and-control regulatory agencies with their rigorous inspection and enforcement programmes. Tough legislations and regulations were passed to regulate pollution with contaminating industries requested to obtain written permission before operating or acquiring polluting facilities.

Unlike many developing countries seeking to boost foreign investment and develop their industrial base, at this same development stage Singapore made sure that regulatory requirements were and continue to be backed up through the ENV's robust monitoring, inspection and enforcement regime. During the early stages of industrialization, the country needed to attract foreign investment in manufacturing in order to promote development, for which standards were gradually tightened up over time and industrial polluters were given time to comply and also offered help, if necessary (Rock, 2002).

The following discussion focuses mainly on the water-related pollution regulatory instruments that have been developed along the years. Other important environmental policy instruments will also be mentioned briefly, especially when they are related to water policy instruments or affect water management in any way. Sea pollution control and law pertaining to the sea have been excluded from the scope of this analysis.

From independence to the Ministry of the Environment

Immediately after independence, an integrated Plan of Action for Environmental Protection was implemented (Ministry of the Environment, 1973). Since in the 1950s and 1960s environmental laws were mostly concerned with the protection of public health, the first phase of the action plan was the reorganization of the Environmental Health Division (EHD), which was primarily responsible for street cleaning at that time. With the establishment of a cohesive chain of command and clearly defined responsibilities, EHD was restructured to achieve greater staff discipline and efficiency.

Until 1963, Singapore was governed by public health related provisions included in the Municipal Ordinance (Cap. 133 of the 1933 edition), first passed in 1913, which traced back its origin to the British Public Health Act of 1897. The relevant provisions included in the 1963 Local Government Integration Ordinance Part IV were no different to those detailed in the Municipal Ordinance, except that they were in conformity with the various changes in the local government made in 1959.

The 1968 Environmental Public Health Act (EPHA), implemented from February 1969, sought to integrate the environmental health responsibilities under the Public Health Division. It also aimed at creating a standard code for health-related issues regarding public cleaning and hawking services, markets, food establishments, general health, sanitation and hygiene. The EPHA was undertaken by the Public Health Authority and the Commissioner of Public Health, both of them working under the direction of the Minister of Health. The EPHA covered almost all fields of environmental health except for air and water pollution. It was modelled on similar legislation from Britain, New Zealand and on the New York City Health Code.

The Act ensured a general increase in the amounts charged for penalties, which not only corresponded to the offence but also served as a deterrent for future infractions and other infractors. For example, the EPHA's Section 26 provided comprehensive provisions against littering and depositing refuse in public places, also setting very high fines of up to $500 for first-time offenders and of $2,000 for subsequent convictions. Stricter penalties were also included for those who did not take adequate precautions during construction operations and thus threaten the life of individuals or damage properties.

Other notable provision was the speedy trial of offenders. Singapore opted for rigorous pollution controls that also included effective enforcement through fast prosecution processes when the offender was found responsible for contravening either the law or the related regulations. For example, under the EPHA, the offender would be fined immediately when contravening Section 26 or 27 of the Act. Under this procedure, the offender was given a ticket on the spot and was required to attend a designated court on a prescribed day. The offender was dealt with summarily if he pleaded guilty. No further action was taken if fines were paid, but if the offender insisted on having a trial, a date was fixed for hearing. If the offender failed to attend the court on the given day, a warrant would be issued to proceed with the arrest.

Regarding sanitation and sewerage, Section 67 of the Act provided that new buildings should include sufficient and adequate sanitary facilities and that these should be properly maintained. Similarly, Section 72 addressed the proper maintenance of sanitary facilities at public places.

In 1968, the passing of the Local Government (Conservancy) Regulations and the introduction of the Environmental Public Health (Public Cleansing) Regulations in 1970 both reinforced the environmental improvement processes. These regulations forced the public to take positive steps to keep the surroundings of their premises clean. It also made the residents or business owners responsible for cleaning the footpaths in front of their premises, including corridors and passageways.

In 1970, with the parliamentary discussion of the Environmental Public Health (Amendment) Bill, the Minister of Health proposed to strive for even higher cleaning standards (Hansard, 1970a). This amendment addressed the problem of garbage thrown into streams, rivers, canals and drains as well as polluting reservoirs, lakes and catchment areas. As such, these acts would no longer be dealt through notices and summons as formerly provided under the cumbersome procedures detailed in Section 50, subsections (2)(c) and (3) (d) to (g), of the 1968 Environmental Public Health Act. This amendment allowed authorities to immediately fine any person contravening any aspect of public cleaning regulations.

Water pollution control

In the early years after independence, water pollution control was very complex from the institutional viewpoint and laws were insufficient and scattered among various regulations. Multiple agencies used to manage water pollution related issues. The PUB (then under the Ministry of Trade and Industry, MTI) was responsible for adequate and reliable industrial and domestic water supply; and the Public Works Department (PWD, under the administrative control of the newly formed Ministry of National Development, MND) was responsible for sewerage and drainage. Sanitation and cleaning services were the responsibility of the Ministry of Health.

The enforcement of regulations on wastewater outlet connections, connection to sewers, adequate sanitary provision and wash areas depended on different authorities. Examples include Regulation 9(1) of the 1965 Local Government Drainage and Sanitary Plumbing Regulations; Section 2(2) and 36(7), (8) and (9) of the Local Government Integration Ordinance; Municipal By-Law 14 (Private Sewage Plants) issues in 1951; and Section 11(1) and (2) of the 1965 Local Government (Drainage and Sanitary Plumbing) Regulations. The question of authority was further complicated with the introduction of the EPHA in 1968, which empowered the Commissioner of Public Health to serve notices to owners to provide adequate sanitary facilities.

In 1969, the then Prime Minister Lee Kuan Yew intervened in the cleaning of highly polluted waterways. He wanted engineers to develop and implement a plan to prevent polluted or sewer water flowing into the island's several waterways as well as to dredge the rivers to keep them clean. He also instructed the MND to make a real effort to clean up the main waterways in Singapore such as the Kallang Basin and the Singapore River (Lee, 1969). Several measures were taken as a result of these requests and detailed reports were prepared (Ling, 1969). For instance, in terms of regulations, those to control trade effluent discharge into sewers and watercourses were listed under the Local Government (Disposal of Trade Effluents) Regulations from 1970 (Tan *et al.*, 2009). In addition, certain provisions in the 1970 Natural Reserves Act were applied to the Central Catchment Area reservoirs. These measures prohibited the introduction of animals into natural reserves, as well as destroying or damaging any objects of zoological, botanical or other scientific or aesthetic interest in the reserves (Science Council of Singapore, 1980).

Despite these efforts, the extent of water pollution in the watercourses was chronic. In 1971, only half of the wastewater produced (approximately 57 Mgal/day) was properly treated, with the remaining 53 Mgal/day contributing to water pollution in Singapore. Some 15 Mgal/day were industrial discharges, 5 Mgal/day came from hawkers and markets, 6 Mgal/day were produced by washing, 12 Mgal/day were sullage water from Kampongs and 10 Mgal/day were produced by government agencies (Ministry of the Environment, 1973).

On 21 July 1971 during the Second Parliament's opening session, the Presidential Address mentioned water pollution as a big problem and urged the Parliament to act upon it: 'All kitchens, sullage, bath and other polluted household waste must go into the sewers and not into open drains. Industrial wastewater must be treated to reduce the pollution before being discharged into canals, open drains or sewers' (Parliamentary Debates, 1971).

Numerous new regulations and amendments to existing laws were passed. For example, in 1972, under Section 88 of the Public Utilities Ordinance from 1963, the PUB (Catchment Area Parks) Regulations made the agency responsible for managing the catchments in the parks. These included MacRitchie Reservoir and Park; Seletar Reservoir; Seletar Zoo; Panadan Reservoir; Kranji Reservoir; Pulao Tekong Reservoir and Upper Peirce Reservoir Park. With these modifications, the bathing or washing of any animal, person or object in the reservoirs was prohibited to avoid polluting them.

Introduced in 1971, the Environmental Public Health (Prohibition of Discharges of Trade Effluents into Water Courses) Regulations and above-mentioned 1970 Local Government (Disposal of Trade Effluents) Regulations, defined standards for various types of chemicals and solid suspended material in discharged trade effluent into any watercourse. They also recommended the installation of pre-treatment plants whenever necessary (Chia, 1978). Unfortunately, the quality criteria for the discharge of trade effluents established in these regulations did not include a discharge quality suitable for raw water supply. Therefore, this discharge had to undergo further treatment to be reused.

The aforementioned regulations became the precursors of the subsequent Water Pollution Control and Drainage Act (WPCDA) issued in 1975. This Act consolidated all of the pollution control provisions detailed in previous regulations and paved the way for comprehensive water pollution control in Singapore.

The Cleary Report and the establishment of the Air Pollution Unit

The Cleary Report, which focused on air pollution, was one among several World Health Organization (WHO) environmental health studies carried out in Singapore. Previous assessments include the Rogus and Maystre's Engineering Feasibility and Management Study of Solid Waste Handling and Disposal; Casanueva's Public Health Engineering; and Tomassi's Report of a Field Visit to Singapore (Cleary, 1970: 1).

As result of a one-month long survey in 1970, the Cleary Report recommended to establish a separate Air Pollution Control Organization within the Public Health Engineering Branch to focus on all engineering and technological aspects of air pollution control for Singapore as a whole. With the possibility of more acute pollution problems arising in the foreseeable future, the report stressed the need for effective pollution control and skilled technical staff, the expansion of pollution monitoring programmes, coordination among various government agencies to avoid any duplication and waste of resources, and the collaboration of pollution control consultants who would oversee potentially polluting industries.

More importantly, Cleary recommended that air pollution related measures were adopted as an integral part of urban planning and that the Senior Public Health Engineer, or his nominee, were invited to the Planning Committee. He also strongly suggested that the Ministry of Health initiated a programme to educate industrial management and the public at large about the need to control air contaminants. The report also proposed air pollution legislation, such as statutory emission limits, for some pollutants and the management of 'scheduled premises' that referred to polluting industries like cement, ceramic or milling works.

The report advised that specific air pollution legislation were formulated as an integral part of the development of an effective air pollution control organization. Yet it 'strongly recommended that the spirit of the enforcement be by persuasion and advice rather than by coercion'. However, considering the reality on the ground, the government pragmatically adopted a strategy of stringent enforcement and vigorous public awareness campaigns. As such, an Anti-Pollution Unit (APU) was established in the Prime Minister's Office. Primarily responsible for enforcing environmental legislation along with the Ministry of Health, it successfully managed Singapore's pollution control until 1999 when it was repealed.

The Clean Air Bill was presented in the Parliament in October 1971 and passed and promulgated on 2 December 1971 (Government Gazette, 1971). The Clean Air (Standards) Regulations were established on 11 January 1971, defining permissible emission levels for various industrial pollutants (Government Gazette, 1972). The Clean Air Act empowered APU to control air pollution emissions coming from industrial and trade premises. The Act was amended several times. The Clean Air (Amendment) Act was promulgated on 18 April 1975 (Government Gazette, 1975); the Clean Air (Standards) Regulations were amended on 27 February 1978 (Government Gazette, 1978) for 'stricter control over the emission of certain air pollutants such as dust, acid gases, chlorine, carbon monoxide, etc.' (APU, 1979: 16). Finally, on 1 May 1980, the Schedule to the Clean Air Act was amended (Government Gazette, 1980). The objective of this amendment was to 'provide for stricter control over those premises being used for the storage of large quantities of toxic or volatile chemicals' (APU, 1981: 18).

On 11 February 1999, the Parliament passed the Environmental Pollution Control Act 1999 (EPC Act), which repealed the Clean Air Act (Government Gazette, 1999). The new EPC Act consolidated the various pollution controls in different Acts into a single Act. It also gave the ENV additional powers to control or prohibit the emission of any particular air pollutant from any industrial or trade premise (Parliamentary Debates, 1999). As air pollution control was included in the EPC Act, the Clean Air Act was no longer necessary and thus it was repealed. The enforcement of the Clean Air Act, together with the work of the APU and other agencies, saw air pollution effectively under control in Singapore.

Enforcement of environmental legislation prior to 1972

Before 1972, the APU and the Ministry of Health were primarily responsible for enforcing environmental legislations including those related to water pollution. Between December 1968 and July 1971, a total number of 29,525 people were prosecuted. The deterrent effect of this enforcement was so important that when the 1970 Prohibition on Smoking in Certain Places Act was brought into force on 1 October 1971, nobody was prosecuted despite the fact that within two months auditoriums in cinemas and theatres were free of smoke (Ministry of the Environment, 1973).

Policy enforcement has been rigorous and several different ways of ensuring compliance with rules, laws, regulations and measures have been adopted. For instance, on-site inspection has been a predominant style in Singapore. Easy-to-measure tests were frequently conducted to evaluate compliance with emission standards, monitoring of sulphur dioxide and dust fallout. In addition, sampling, emission and effluent source testing, complaint follow-up and equipment inspections were regularly done. It was a general policy that inspections were unannounced and fairly informal unless the firm was polluting. In cases where indications that anti-contamination measures had been breached, inspectors were likely to be stricter and firmer in their evaluations and assessments.

Creation of the ENV

Both the 'Keep Singapore Clean' campaign put in place in 1969 (strategies on public education and information are analyzed in Chapter 5) and the Environmental Public Health Act from 1968 helped improve the quality of Singapore's urban environment and catalysed its transformation into a 'Garden City' (Ministry of the Environment, 1973). This 'Garden approach' and visionary concept of sustainable development merged environment with development, and translated it into a sustained effort to green the growing metropolis. Open forested and gardened spaces sent potential investors a subtle message of effectiveness and efficiency whilst brightening the urban landscape to improve individual, community and social conviviality and interaction.

Almost at the same time, several countries around the world created ministries, departments or other bodies directly in charge of environmental activities. In fact, by 1972, 15 countries (Australia, Canada, Chile, Finland, France, West Germany, Greece, Japan, Malta, New Zealand, Portugal, Senegal, Sweden, the United Kingdom and the United States) had established environment ministries (Caponera, 2003). Immediately after the 1972 United Nations Conference on the Human Environment in Stockholm and following the trend among developed countries, Singapore created the ENV to craft and sustain a good natural environment for its people. To concentrate all environment-related affairs in this newly created body, those departments under the ministries of Health and National Development that had until then dealt with pollution control, sewerage, drainage and environmental health became part of the ENV. Moreover, the Sewerage Department was given the responsibility to control water pollution.

In the 1970s, two types of preventive measures were used to protect water quality. The first one was the control of polluting activities in and around the reservoirs, a task under the PUB's responsibility. The second one, overseen by the ENV's Sewerage Department, consisted of controlling effluents (both domestic and trade) going into watercourses in the unprotected catchment areas. Throughout this time, the APU continued to implement measures to manage air pollution. In 1985, the APU took on the additional responsibility of keeping hazardous substances in check, whilst the ENV implemented new programmes to dispose of toxic waste. These air and water control pollution activities as well as the management of hazardous substances and toxic waste were subsequently placed under the Pollution Control Department formed in 1986 as part of the ENV.

Sewerage management

In 1972, the ENV's Sewerage Department launched a programme to develop a comprehensive sewerage system for all new industrial, housing and commercial developments in Singapore built from that moment onwards. Through a network of underground sewers, pumping stations and sewage treatment works, this scheme ensured the discharge of all wastewater into sewers and then to sewage treatment works for treatment. By 1992, a network of 2,300 km of sewer lines and six major sewage treatment works had been built, treating more than one million cubic meters per day of sewage (Ministry of the Environment, 1993a). Irrespective of whether the discharge went to the sea or to inland waters, the standards for treated effluents were 50 mg/l for both Biochemical Oxygen Demand (BOD) and Total Suspended Solids (TSS).

Overall, factories were encouraged to adopt good housekeeping procedures, follow less polluting processes and use recycled water in their premises to save water and reduce effluents volume. Officers of the Sewerage Department made regular inspections to industrial and commercial premises to ensure that preventive measures were taken to minimize the discharge of waterborne pollutants. For example, factories that generated large quantities of acidic effluents were required to install pH monitoring and recording instruments to monitor the pH of the effluent discharged. These inspections and sample tests led to the prosecution of industries found discharging acidic effluent into the sewers. In 1993 alone, 36 such industries were prosecuted (Ministry of the Environment, 1993a).

For industries not able to treat their effluents to meet the required standards, the Trade Effluent Regulations were amended in 1977 to make provisions for the measurement of BOD and TSS so that a payment could be made for the treatment of those non-complying effluents. The treatment fee was computed on the basis of BOD and TSS concentrations and discharge volumes so that levies were proportional to the pollution load. Subsequently, the tariff's schedule was revised in 1983, 1992 and 1993 to reflect the market cost of wastewater treatment (Ministry of the Environment, 1993a).

In a sustained effort to control pollution, the Sewerage Department engaged several consultancy firms to design and install chemical treatment units, activated sludge plants, trickling filters and sedimentation units. The emphasis on effluent control encouraged many reputed firms to establish wastewater management related businesses in Singapore including consultancy services, equipment supply or laboratories to treat trade effluents, processing plant design, as well as analytical services to monitor trade effluents (Science Council of Singapore, 1980).

Regulatory measures for pollution control were financially backstopped by the government. For example, in the 1980s, Singapore invested annually around 1 per cent of its GDP on environmental management. In 1989 alone, $27.76 million was spent on maintaining and developing greenery. In early 1990s, large sums were poured into implementing the vision of a clean and green city. That decade, the government allocated $609.3 million annually to address environmental issues, including $246.12 million for environmental health, $86.45 million for sewerage schemes and $55.08 for drainage programmes (EDB, 1991).

Water legislation from 1972 to 1992

In the following sections are discussed the several instruments that were developed in the 20 years that followed the creation of the ENV.

Water Pollution Control and Drainage Act 1975

As effective as it was in its time, the Environmental Public Health Act from 1968 touched only on some of the many aspects related to water pollution control (mainly parts VIII and IX). The country's political leadership realized that these provisions were no longer appropriate to face the pollution challenge the growing city-state was facing.

After the then Prime Minister Lee Kuan Yew emphasized the importance of taking urgent actions to clean the waterways, two trade effluent-related regulations were passed: the Local Government (Disposal of Trade Effluents) Regulations from 1970 and the Environmental Public Health (Prohibition of Discharges of Trade Effluents into Water Courses) Regulations issued in 1971. Later on, the need for more comprehensive legislation on water pollution control was felt and the 1975 Water Pollution Control and Drainage Act was enacted.

When presenting the bill on Water Pollution Control and Drainage in the Parliament, the Minister for Law and the Environment, E.W. Barker, outlined the need to enact new laws that would respond to the increasing environmental problems resulting from rapid population growth, as well as from economic and industrial development. He also stressed the growing demand for water and the importance of a cleaner environment amidst an increasingly urbanized environment. He informed the House that the control of water pollution and the provision of effective sewerage infrastructure were as important as ensuring water supply. Stating from the premise that sewerage management was the most effective method for water pollution control, the Minister emphasized the need to pass new laws since the old ones were now insufficient and obsolete (Hansard, 1975).

Within three years of its formation, the ENV formulated the 1975 Water Pollution Control and Drainage Act. This Act was primarily based on the principle that effluents must be discharged into sewers wherever possible and that regulations must focus on effluent quality. According to several criteria, treated discharges had to meet 20/30 standards that indicated a maximum permitted limit of 20 mg/l of BOD and 30 mg/l of TSS in the discharge.

Under this Act, the discharge of any toxic substance into any inland water that could result in environmental hazards was considered a punishable offence. The regulation did not require proof of fault; it was an offence of strict liability and high fines. A maximum penalty of $5,000 was prescribed in Regulation 31(5), Part V, should the law be violated. In 1985, the fine was increased to $10,000 or six months' imprisonment. For subsequent offences, the maximum prescribed fine was of $20,000 and the time of imprisonment could oscillate between one and 12 months. With such strict measures, minimal prosecutions were presented and, in most of the cases, defendants pleaded guilty.

This Water Pollution Control and Drainage Act (Section 2) empowered the Director of the Water Pollution Control and Drainage (an area of the Sewerage Department), to control the discharge of wastewater from all sources. The Act clearly spelled out that no person would be allowed to discharge any trade effluent into a watercourse without the Director's written approval. This newly acquired authority made water pollution control and effluent management regulations more effective as the Director could now revoke or suspend previously granted permits as well as to change the permitted effluent volume, quantity or discharge rate.

The Act's requirements were later reiterated, modified and expanded in a related piece of legislation, the 1976 Trade Effluent Regulations. The new set of laws prescribed for the minimum quality standards to be met by any effluent discharged into controlled watercourses, other watercourses and public sewers as well as for the documentation that was required were discharges to be made into watercourses or public sewers (Trade Effluent Regulations 1976). They also provided for discharge conditional consent, which depended on the submission of details of the proposed operation, water consumption and the nature of the effluent. The changes that would affect the effluent or nature of the effluent had to be notified within 14 days of such change and, in all cases, trade effluents had to be treated prior to discharge. Additionally, the 1976 Trade Effluent Regulations dealt in detail with the monitoring facilities, equipment and particulars of operational discharges.

Similarly to all other environment related laws, enforcement was strictly followed. For example, a company engaged in the manufacture of printed circuit boards was charged for discharging concentrated copper and ammonia without the written permission of the Director of the Water Pollution Control and Drainage Department. The company pleaded guilty and was fined $750. Unfortunately, however, the effluent caused a breakdown of the biological treatment process at the Kranji Treatment Works, for which the ENV claimed $52,300 in cleaning and administrative costs. Initially, the company denied any responsibility, but eventually paid the entire amount (Foo, 1993).

The Trade Effluent Regulations of 1976 prohibited certain discharges and controlled downstream water quality by setting parameters on the range of effluents that could be discharged into a watercourse. The spirit of the legislation does not 'grant a right to pollute', but ensures that water entering the watercourses is of a certain quality. As such, parameters varied on controlled watercourses, other watercourses and public sewers. Unsurprisingly, the most stringent criteria were applied to controlled watercourses because of their importance for drinking water supplies. For similar reasons, trade effluent discharges were prohibited in the reservoirs of the central catchment area.

Initially, the regulation's strict and rigorous standards became a matter of concern for some industries, especially using or producing secondarily concentrated organic waste when they did not have means to install expensive treatment plants. The regulations were consequently amended to provide technical support for these industries.

Other acts and regulations

The government's basic approach to trade effluent control was to deal with pollutants at their source through preventive and palliative measures (Hansard,

1977). Alongside effective law enforcement, and as basic infrastructure, agencies extended effective and quick sewerage extension to all premises, which ensured that wastewater went into sewers and not into open drains. Efforts also included revising already existing laws to adapt to changing realities in the island.

After the 1976 Regulations, other measures were taken to try to devise alternatives for those polluting industries that were not able to treat their effluent according to the stipulated standards, namely the 1977 Trade Effluent (Amendment) Regulations and the Sewage Treatment Plants (Amendment) Regulations from 1978. These new sets of laws were effective in addressing the concerns of those non-complying industries, as the ENV was to treat their effluent at an expense the industries would cover. The Amended Trade Effluent Regulations furthered permitted the Director of Water Control and Drainage to allow polluters to discharge effluents with concentrations in excess of the BOD and TSS standards. Since these regulations included an annually reviewable trade effluent tariff scheme, the fee charged to the industries was calculated on an increasing scale according to the concentration of BOD and TSS. As the ENV was treating the effluents for a fee, the scheme did not make any changes in terms of environmental pollution and water quality. In fact, it merely shifted the responsibility of effluent treatment from individual industries to sewage treatment plants. Furthermore, the fee scale was such that at very high concentrations, industries found it economical to install their own treatment plants instead of paying the ENV to treat their discharges (based on Trade Effluent Tariff Scheme included in Appendix 8 of the 1992 Pollution Control Report (PCD, 1993: 3)). The 1976 Trade Effluent Regulations and subsequent amendments on the tariff scheme meant these measures were in complete accordance with the polluter pays principle (McLoughlin and Bellinger, 1993).

Additionally, laws were also passed addressing technical issues. Once the 1975 Water Pollution Control and Drainage Act was issued, the 1975 Sanitary Appliances and Water Charges Regulations and the 1976 Sanitary Plumbing and Drainage System Regulations followed. These rules were mainly technical in nature and were issued with an objective in mind to minimize leaks, pollution and water wastage.

The government was prompt to enact new laws or amend old ones as and when they considered it necessary. For example, in April 1983, a substantial amount of toxic waste was found in the Kranji Catchment. The waste, containing mostly copper chloride, had leaked through the ground and into a stream that flowed into the Kranji Catchment. Fish in the water died within few hours. As a reactive and deterring measure, the government put in place strong penalties to discourage the indiscriminate discharge of toxic waste (Hansard, 1983). Consequently, an amendment was introduced in the 1975 Water Pollution Control and Drainage Act. With the insertion of the new section 14A, these types of discharges were made a punishable offence liable to a penalty of up to $10,000, six months' imprisonment, or both for a first offence. Subsequent offenders could be imprisoned and fined up to $20,000.

Pollution Control Department

Since its inception in 1986, the Pollution Control Department administers all acts and regulations related to water pollution (PCD, 1994). The only exception at that time applied to laws regulating the discharge of solid waste into water bodies, a competency included under the Environmental Public Health Act from 1968 and therefore the responsibility of the Ministry of Health.

The Pollution Control Department has the overall responsibility for the control of air, water and noise pollution, hazardous substances and toxic waste. It thus monitors air, inland and coastal water quality as well as formulates and implements joint programmes on transboundary pollution with neighbouring countries.[3] It also works with several other agencies at the planning and building stages of new developments to incorporate anti-pollution measures from the very design phase itself.

Industrial land-use zoning

Zoning is a planning tool adopted comprehensively in Singapore for both micro and macro developments. It regulates the location of various land uses to ensure that similar and compatible uses are located together and conflicting ones are located as far away as possible. For example, the industrial estates in Jurong, Sungei Kadut/Kranji, Woodlands and Sembawang are situated along coastal lines, removed from densely populated areas. Heavy industries are located nearest to the coast, followed by medium and light industries established away from the sea. Within the different zones, sub-zones have been created for specific industrial categories such as foodstuffs, chemicals, etc. Large-scale polluting industries, for example oil refineries, are not located in the mainland and, instead, are based in the Pulau Bukom, Pulau Merlimau, Pulau Pesak, Pualu Ayer Chawan and Pulau Ayer Merabu islands.

Subsequently, land use planning has played a very important role in protecting Singapore's water catchments. For instance, in order to control developments within the unprotected catchment areas, the Water Catchment Policy of 1983 established that a maximum 34.1 per cent of land could be developed within the water catchments, limiting population density to 198 dwelling units per hectare until 2005. Together with stern pollution control measures, this restriction ensured well-planned urban development and good water quality even from unprotected water catchments (Tan *et al.*, 2009).

Water pollution control – phase 2

During the 1972–1992 period, water pollution was reduced to WHO accepted levels. Inspections and prompt actions on complaints played an important role in this improvement. For example, from 1986 until 1992, there was an average of 140 complaints per year in relation to pollution by chemicals, oil, industrial wastewater, domestic water, etc. The highest number of complaints reached 171 in 1989 whilst the lowest number totalled 81 in 1986. Regarding inspections, the Pollution Control Department conducted regular trade effluent tests. Data available for the same period shows that, on an average, 2,425 tests were conducted yearly and an annual average of 310 failures were detected (Ministry of the Environment, 1986–1993). The observed trend shows the number of tests corresponded with the number of complaints; thus, to a higher number of queries, more inspections followed.

Those years, 1972–1992, signalled a period of transformation for Singapore. As the city-state embarked relentlessly on an industrialization and urbanization process, legislation, development of infrastructure and delivery of services played a leading role in meeting air and water quality standards well within the acceptable ranges prescribed by both the US Environment Protection Agency and the WHO. Over the course of these years, Singapore drafted its first Water Master Plan (see Chapter 1) and Water Conservation Plan, the first one dating back to 1981; engaged in flood alleviation efforts in urban areas; built incineration plants (Ulu Pandan in 1979, Tuas in 1986 and Senoko in 1992), estuarine reservoirs, water reclamation plants (Jurong in 1974 and 1981, Bedok in 1979, Kranji in 1980 and Seletar in 1981); eradicated malaria in 1982; and by 1987 had both successfully completed the cleaning of its rivers and rooted out the use of nightsoil buckets (Tan *et al.*, 2009). All these attainments would not have been possible without comprehensive legislation and its effective implementation.

A nation with very small area (714.3 km^2 in 2011), very high population density, one of the largest ports in the world, several oil refining centres and a myriad of industrial estates is bound to be pollution prone, especially when it is also a hub for chemical, pharmaceutical industries and electronics. A comparison between the highly polluted Singapore of the 1960s with the Singapore of 1992 shows remarkable achievements in terms of clean land, air and waters. With a synergic vision, appropriate policies, responsive programmes, rigorous legislation and the encouragement of inter-institutional coordination to deal with pollution, the government fulfilled the promise it made in Parliament in 1968 to transform Singapore into a clean, green Garden City.

A model green city

In 1992, a master plan was developed to prepare the island for the next decade: the Singapore Green Plan for Environmental Protection and Improvement. The same year, it was presented at the United Nations Conference on Environment and Development in Rio de Janeiro.

The Green Plan was formulated with inputs from various government agencies, ministries, private organizations and members of the public. It mapped the policies and strategies to transform Singapore into a model green city with high public health standards and an environment conducive to leading a high quality of life (Ministry of the Environment, 1993b). The Plan was subsequently reviewed and actualized in 2002 and 2005 to ensure it remained relevant. An updated edition, the Singapore Green Plan 2012, was published in February 2006, setting out the broad directions and strategic thrusts that will help ensure Singapore's long-term environmental sustainability (MEWR, 2006).[4]

Since independence, the government had grown aware of the importance of further involving the population in the protection and conservation of the limited water resources available. It thus opened reservoirs to the public for non-polluting recreational activities (Hansard, 1989a) (for detailed analyzes, see Chapter 5). With remodelled institutions and updated legislations, water management in Singapore entered a new era.

Littering in Singapore had been reduced to a large extent due to sustained efforts but it was never completely eradicated. For course correction, the Singapore government introduced the Corrective Work Order (CWO) (EPHA 1992 Amendment), which is still in force. According to this regulation, instead of hefty fines, offenders are required to clean up their community and public areas, under full sight of community members, for several periods of up to three hours each, subject to a total of 12 hours. This applies to serious littering crimes and repeat offenders aged 16 years and above. It is thought that other than punitive, these measures will have a reparative effect forcing the offender to experience the difficulties cleaners face. Despite being a controversial and unpopular choice, the government adhered to the Corrective Work Order, showing its commitment to a cleaner environment.

Code of Practice on Pollution Control

First published in 1994, the Code of Practice on Pollution Control (CPOOC) is based on the principle that environmental impacts caused by infrastructural developments and related pollution problems can be prevented or mitigated by placing such constructions in designated areas and incorporating pollution control measures as these are designed and from the planning stage (PCD, 1997). The code established that locating requirements for new developments and their compatibility with the surrounding land should be planned and implemented by both the planning and developing authorities, and in consultation with the Pollution Control Department. For these zoning exercises, the Code provides elaborate guidelines and requirements in its three main sections. Part 1 includes the general requirements to locate industries and for the submission of development proposals for planning approval. Part 2 lists technical and specific requirements on pollution control for building plan submission, and part 3 outlines guidelines when applying for the licences and permits needed to operate a factory.

The CPOOC classifies industries under four categories, namely: clean, light, general and special. For clean industries such as IT, fashion designing or manufacture of paper products without printing activities, no buffer distance from residential building is required. They may even be in a water catchment provided that the premises are connected to public sewers. Conversely, a buffer distance of at least 50 m needs to exists between light industries and the nearest residential building. These industries, comprising activities as diverse as biotechnology; footwear and plastic products manufacturing; and printing and publishing activities are thus not allowed to generate large quantities of trade effluent or solid waste.

For general industries, a buffer distance of 100 m is prescribed. As specified by the Pollution Control Department, the manufacture of electronic appliances and cutlery hand tools; cutting, grinding and polishing marble and ceramic tiles; as well as vehicle repairing and servicing etc., have to install, operate and maintain pollution control equipment to minimize air, water and noise pollution from their operations. Lastly, for special industries the buffer distance is kept at 500 m and also required to install, operate and maintain pollution control equipment. If any of them is involved in the use or storage of large quantities of hazardous substances, they may be required to conduct an impact assessment to support their application in order to be located in the appropriate designated area. In addition, a compulsory hazard analysis is required to establish health and safety zones and prevent any knock-on effects from the neighbouring hazardous installations and to protect the public from any danger of fire. Some examples of these special industries include those dealing with dairy products, alcohol, paints, varnishes, canning and preserving fruit and vegetables, iron-steel basic industries, tanneries, saw mills, pharmaceuticals, as well as those manufacturing tyres, tubes and motor vehicles.

Furthermore, Section 2.1.2 of the Code of Practice on Pollution Control provides specific requirements regarding used water from residential, commercial and industrial buildings. Industries that do not meet the standards stipulated in the Trade Effluent Regulations from 1976 need to install, operate and maintain a treatment plant for their effluent to meet the required standards before discharge. The CPOOC provides detailed guidelines to design and operate a production area, washing facilities, loading/unloading areas, cooling towers, compressor or generator rooms, boiler rooms, stores, storage tanks, chemical warehouses, buildings and laboratories. Similarly, it does not permit the dilution of trade effluent with potable, rain or industrial water to comply with the standards. The Code also prescribes similar effluent discharge limits from the wastewater generated by aquaculture farming. It does not allow livestock to be reared in the open and sets guidelines for solid waste and wastewater produced by animal rearing and animal farms.

To avoid the pollution of rainwater, the Code establishes that rainfall shall not be discharged into a public sewer but channelled into a watercourse. Following this same logic, contaminated rainwater from process areas is to be collected and treated before discharge into the watercourse. Further information is included in the CPOOC's various appendices, which provide a List of Toxic Industrial Wastes, Hazardous Substances, Trade Effluent Discharge Limits, Trade Effluent Tariff Scheme and online pH Monitoring and Effluent Discharge Control, etc.

Water Catchment Policy 1999

As discussed earlier, the Water Catchment Policy of 1983 ensured planned urban development could take place without sacrificing water quality in unprotected urban catchments. In the 1990s, by adopting advanced water treatment technology in the upgrade of treatment plants, the PUB was able to use water from increasingly urbanized and unprotected catchments. This made it possible for the government to lift the urbanization cap and population density limit within the unprotected water catchments. This demonstrated how planners and various agencies could work together to review and improve a piece of public policy so that it reflects the evolution of technological advances and pollution management practices (Tan *et al.*, 2009). Policy evolution has also extended to land use in non-catchment areas. In Singapore, land zoning has not been a one-time or stationary practice. In fact, land has been re-zoned and sites redeveloped to turn such spaces into higher-value areas.

Repeal of the Water Pollution Control and Drainage Act

In 1999, the Water Pollution Control and Drainage Act from 1975 (Chapter 394) was repealed and relevant powers streamlined into the Sewerage and Drainage Act and the Environmental Pollution Control Act. Both of the Acts were issued that same year and included stipulations on the control of discharges into public sewers and watercourses respectively. This was a second logical step towards making the PUB Singapore's national water agency within a broader environmental framework overseen by the ENV. Subsequently, in 2002, the implementation and enforcement of the Environmental Pollution Control Act became the NEA's responsibility.

The 1999 Sewerage and Drainage Act (SDA) compiles the relevant rules on the provision, operation and maintenance of the sewerage system. This Act stipulates that all used water is discharged into public sewers, whenever they are available, and manages the treatment and discharge of industrial wastewater into public sewers. Penalties for breaching any of the different provisions are also listed in this Act and in various regulations made under it. The competency of overseeing compliance with the Sewerage and Drainage (Trade Effluent) Regulations was transferred to the PUB when the Board was reconstituted as the national water agency in 2001. As such, this Sewerage and Drainage Act allowed the PUB to control all types of used water, and in addition to its many functions related to water supply, the Board was given the institutional responsibility to maintain and manage public sewerage systems, public sewers and storm water drainage systems, drains and drainage reserves. The Act and its related regulations will be discussed subsequently.

Fast-paced infrastructural developments initiated by the Revised Concept Plan of 1991 created the need for the effective maintenance of the entire sewerage system. In March 2000, the first edition of the Code of Practice on Sewerage and Sanitary Works was published by the Sewerage Department, which on 1 October 2004 was reorganized as the PUB's Water Reclamation (Network) Department. This Code replaced the 1976 Code of Practice of Sanitary Plumbing and Drainage System, and the 1968 Sewerage Procedures and Requirements.[5]

The Code was issued under Section 33 of the Sewerage and Drainage Act of 1999. It aimed to be a guide for qualified practitioners planning and designing

sanitary and sewage systems. As such, it provides the minimum and mandatory design requirements for such systems, and includes some good engineering
practices.

In March 2000, the first edition of the Code of Practice on Surface Water
Drainage was published by the Drainage Department. This Code specified the
minimum engineering requirements for surface water drainage for new developments. Its objective was to provide an effective and adequate drainage system to
prevent flood and public health risks.

Integration of water administration

The Board was consequently transferred from the Ministry of Trade and Industry
to the ENV, retaining its water supply portfolio and absorbing ENV's sewerage
and drainage departments. At that same time, the regulation of electricity and gas
industries was transferred to a new statutory board, the Energy Market Authority
(further analyzes on this issue can be found in other chapters in this book).

In the process of transforming the PUB into the national water agency, the
Board was reconstituted under Section 3 of the 2001 Public Utilities Board Act
(Attorney General's Chambers).[6] The enactment of the 2002 Statutory Corporations (Capital Contribution) Act no. 2 (No. 34 of Schedule of Statutory Corporations – Capital Contribution – Act, 2002. Act no. 5 of 2002) and related
amendments in the PUB's 2001 Act made the Board responsible for water reclamation and for the management and maintenance of the drainage system, in
addition to ensuring the supply of piped water for human consumption. Part V of
the 2001 Public Utilities Board Act, which deals with offences, imposes various
responsibilities on both the PUB as well as on the public. Section 45 makes staff
responsible for the maintenance, keeping and safety of water installations.

Section 50 of the 2001 Public Utilities Board Act deals with unauthorized connections, water contamination and waste of water. For example, a person with an
unauthorized connection or a person found wasting or contaminating water or
interfering with water supplies shall be liable to a fine of up to $50,000 or to an
imprisonment term of up to three years or both. Any recidivism shall result in a
daily fine of $1,000 or part thereof, if the offence continues after conviction.

In 2002, the ENV's Environmental Public Health and the Environment Policy
and Management divisions merged with the Meteorological Service Department,
part of the Ministry of Transport, to give birth to the National Environment
Agency (NEA), also under the ENV. This statutory board position gave the
NEA greater administrative autonomy and flexibility to assume its responsibilities properly and promptly. It also made it the agency responsible for enforcing
the 1999 Environmental Pollution Control Act, the Hazardous Waste (Control
of Export, Import and Transit) Act, the 1999 Environmental Public Health Act
and the Control of Vectors and Pesticides Act.

Within the NEA, the Environmental Protection Division was formed to create
and maintain a quality living environment, control pollution, as well as to ensure
sustainable development and resource conservation. This Division is accord-

ingly divided into four departments in charge of Pollution Control, Planning and Development, Resource Conservation and Water Management. Among these, the Pollution Control Department is entrusted with the responsibilities of air, water and noise pollution control as well as for the control of hazardous substances and toxic waste.

In 2004, the ENV was renamed Ministry of Environment and Water Resources (MEWR). With these institutional reforms and under one ministry, a broader focus and more comprehensive environment and water administration was achieved. Under this leaner, more streamlined and more policy-focused ministry, effective policy implementation was left to two statutory boards: the PUB and NEA. This new structure has successfully added an environmental dimension to water resources administration.

The Environmental Pollution Control Act of 1999

The Environmental Pollution Control Act (EPCA) passed in 1999 came into existence after the consolidation of previous and separate laws on air, water and noise pollution and on the control of hazardous substances. The EPCA replaced the Clean Air Act (Chapter 45) and Part VI of the Environment Public Health Act from 1968 (Chapter 95), and consequentially caused amendments in the Poisons Act (Chapter 234) and the Medicine (Advertisement and Sale) Act (Chapter 177) (Hansard, 1999).

It was expected that with the continuous growth and expansion of Singapore's industrial sector, the amount of pollutants would consequently increase and therefore adequate and appropriate regulatory powers would be required to keep it under control. It was also expected that the new set of standards for emitting or discharging pollutants would instil discipline and promote self-regulation in both industries and the general public. In answering a question related to imposing stringent control standards, and while tabling the bill in the Parliament, the ENV affirmed that:

> whenever a project proposal is received by the EDB and the various government departments that has a pollution element in it, that project is referred to us and we will work with the company very early in the project planning stage to make sure that they understand what are our requirements and that they build that capability to achieve our requirements into the process, because that is the most cost effective way. So our requirements become an integral part of their routine process, which is why despite our very high requirements, I think no company has come back to tell us that our standards are not achievable.
>
> (Hansard, 1999: column 2079)

As a rather broad document, Part V of the Act deals with Water Pollution Control. A number of sections authorize the Director-General of Environment Protection to exert a certain degree of discretionary measures. For instance, Section

15 allows him to issue licences for the discharge of trade effluent, oil, chemical, sewage or other polluting matters and Section 16 allows him to grant licences for the construction of trade effluent treatment plants. Section 17 lists penalties for discharging toxic substances or hazardous substances into inland waters and Section 18 allows the Director-General to take measures to prevent water pollution.

This 1999 Environmental Pollution Control Act was amended and renamed the Environmental Protection and Management Act (EPMA) effective from 1 January 2008. Far from being symbolic, the Act's name change reflects its expanded scope to cover environmental protection and management and resource conservation. This widened range of responsibilities included the energy-labelling of electric appliances, the insertion of Part XA on Energy Conservation immediately after Section 40 8, as well as by amending Section 72 (dealing with offences) to raise the maximum composition fine starting from 1 October 2007 (NEA, 2008). The Act now provides a comprehensive legislative framework for environmental pollution control and the promotion of resource conservation (Tan *et al.*, 2009).

The discharge of wastewater into open drains, canals and rivers is regulated by the Environmental Protection and Management Act of 2008 and the Environmental Protection and Management (Trade Effluent) Regulations. The EPMA and its Regulations are then administered by the Pollution Control Department of NEA's Environmental Protection Division within the NEA.[7]

The Environmental Public Health Act (Chapter 95)

This Act has come to its present form after amendments were made to the original 1987 Environment Public Health Act in 1989, 1992, 1996 and 1999. Later, as a consequence of changes made to several other Acts, the necessary amendments were introduced in 2002 during the enactment of both the National Environment Agency Act and the Sale of Food (Amendment) Act.[8]

As what pollutes land ultimately pollutes water, the Environmental Public Health Act (Chapter 95) is key for water pollution control. In its Part III, the Act deals with public hygiene and Part IV with food establishments, markets and hawkers. Part VI addresses unsanitary premises, sanitary facilities, drains, sewers and wells. Sections 5 and 7 make the Director-General of Public Health responsible for arranging for the cleanliness of public streets as well as for the instalment of any number of dustbins to get the refuse deposited. After an amendment in 2002, Section 6 (3) allows the Director-General of Public Health to require any person (owner or occupier of premises or private streets to which they have access) to sweep, clean and water the adjacent street and to collect and remove any generated garbage. Section 78 (Part IX) of the Environmental Public Health Act (Chapter 95) prohibits selling or offering polluted or unwholesome water. Section 80 makes the NEA responsible for the development of regulations to prescribe standards relating to the quality of water supplied in any area or premise.

Apart from the Environmental Public Health Act (Chapter 95), two other NEA administered and enforced regulations are relevant in administering water. First, the Environmental Protection and Management (Trade Effluent) Regulations which grant the Agency the authority to regulate and set standards for trade effluent discharged into any watercourse, and also the discharge of toxic or hazardous substances into inland waters. Second, the Environmental Public Health (Quality of Piped Drinking Water) Regulations which establishes that as the public health authority, the NEA can regulate and set standards for the quality of piped drinking water. Based on WHO Guidelines for Drinking Water Quality (2006), these regulations also require that piped drinking water suppliers develop a water safety and a monitoring plan. This reflects the high importance legislatures attach to the quality of piped drinking water in Singapore as a separate agency has been put in charge of monitoring potable water quality. This reinforces the credibility of good water governance in Singapore.

The Environmental Public Health (Toxic Industrial Waste) Regulations were also reviewed and amended in 2009 to reflect the rise of new industries, as well as emerging environmental issues and changing policies. Onerous legislative procedures that were no longer valid or relevant in today's context have been simplified. Additionally, and to enhance enforcement, penalties for less severe offences were reduced to fines instead of court prosecution (NEA, 2009).

The Sewerage and Drainage Act (Chapter 294)

The Sewerage and Drainage Act (Chapter 294) (Act No. 10 of 1999) compiles a comprehensive set of laws on sewerage (Part III); drainage (Part IV); protection of water resources (Part V); registration, codes of practice and certificates or approval for works (Part VI); and enforcement (Part VII) and other related issues. As mentioned earlier, used water from all sources is required to be discharged into the public sewer. Nevertheless, if a public sewer is not available, wastewater may be discharged into watercourses (canals and drains) upon meeting much more stringent standards and only after the approval of the relevant authorities.[9]

Section 16 (1) of this Act (Chapter 294) prescribes a maximum penalty of $20,000 if trade effluents are discharged into public sewers without the Board's written approval. If any person is found guilty of discharging any effluent and/or substance into public sewers, he is to immediately terminate such discharge and cease the activity producing such effluent. Section 17 mentions a fine of up to $40,000 or imprisonment for up to three months or both if a person fails to comply.

Moreover, if damage is caused to public sewers or drain lines by any act, Section 19 and 20 lists the corresponding punishment and penalties, which are similar to those mentioned in Section 16. Sections 31 to 35 also details penalties for offences against drains and the drainage system; for intercepting water from any place or the sea within the territories of Singapore; for neglect by qualified persons; and for carrying out any work without a clearance certificate whenever this is needed, etc.

Regulations made under the Sewerage and Drainage Act (Chapter 294) also state that sanitary facilities are to be available in all premises; it addresses the provision of small sewage treatment plants and the control of trade effluent discharged into the sewerage system; touches on the drainage system and sanitary appliances; and explains application fees and water charges levied to maintain the public sewerage system. These regulations are as follows:

1. Sewerage and Drainage (Composition of Offences) Regulations.
2. Sewerage and Drainage (Trade Effluent) Regulations.
3. Sewerage and Drainage (Application Fee) Regulations.
4. Sewerage and Drainage (Sanitary Appliances and Water Charges) Regulations.
5. Sewerage and Drainage (Sanitary Works) Regulations.
6. Sewerage and Drainage (Sewage Treatment Plants) Regulations.
7. Sewerage and Drainage (Surface Water Drainage) Regulations.

Recent trends in water pollution

An analysis of Singapore's water pollution data from the past eight years shows remarkable achievements in water pollution control. This is analyzed further in the following sections.

Solid waste disposal and recycling

Changing consumer habits, lifestyles, packaging requirements and the disposable nature of many consumer goods has led to steady increase in the volume of waste disposed of annually. Now, with the fully operational Semakau offshore landfill, Singapore has built for itself a breathing space for the next couple of decades.[10] Though the management of solid and land waste disposal still remains a daily challenge, so far, the appropriate update of regulations and their efficient enforcement as well as the island's excellent infrastructure have ensured that services are both efficient and effective. Despite the non-uniform increase in waste output, Singapore has been able to enlarge its waste recycling rate (see Table 3.1a and 3.1b).

Table 3.1a Waste statistics and recycle rate

	2003	2004	2005	2006
Total waste disposed (tonnes)	2,505,000	2,484,600	2,548,800	2,563,600
Total waste recycled (tonnes)	2,223,200	2,307,100	2,469,400	2,656,900
Total waste output (tonnes)	4,728,200	4,789,700	5,018,200	5,220,500
Recycling rate (%)	47	48	49	51

Table 3.1b Waste statistics and recycle rate

	2007	2008	2009	2010	2011
Total waste disposed (tonnes)	2,566,800	2,627,600	2,628900	2,759500	2,859500
Total waste recycled (tonnes)	3,034800	3,342,600	3,485200	3,757500	4,038800
Total waste output (tonnes)	5,601600	5,970,200	6,114100	6,5170	6,898300
Recycling rate (%)	54	56	57	58	59

Source: Annual Reports, National Environment Agency, Singapore.

Monitoring of inland waters

Water quality in those reservoirs located within the water catchment areas is jointly monitored by the Pollution Control Department and PUB. Together, these agencies make quarterly assessments of more than 34 streams and 14 ponds in the water catchment areas.

Tables 3.2a and 3.2b compile Singapore's water quality for the past eight years, based on the measured levels of Dissolved Oxygen (DO), BOD and TSS. The data is shown as a percentage of standard quality samples to the total samples checked. Similarly, water quality in non-water catchment areas is also presented. Measurements are collected quarterly from 20 rivers and streams in non-water catchment areas and from what can be observed water quality in both catchment and non-catchment streams remained consistently good from 2002 to 2010.

Table 3.2a Monitoring results of inland waters

Parameters monitored		2002	2003	2004	2005
DO (>2 mg/l)	Water catchments streams (% of time)	100	100	98	100
	Non-water catchments rivers/streams (% of time)	91	95	97	95
BOD (<10 mg/l)	Water catchment streams (% of time)	91	91	92	95
	Non-water catchments rivers/streams (% of time)	91	90	91	92
TSS (<200 mg/l)	Water catchment streams (% of Time)	95	97	100	98
	Non-water catchments rivers/streams (% of Time)	100	96	99	100

Source: Annual Reports, National Environment Agency (NEA), Singapore.

Table 3.2b Monitoring results of inland waters

Parameters monitored		2006	2007	2008	2009	2010
DO (>2 mg/l)	Water catchment streams (% of time)	100	100	100	100	100
	Non-water catchments rivers/streams (% of time)	94	96	92	96	100
BOD (<10 mg/l)	Water catchment streams (% of time)	100	99	99	97	99
	Non-water catchments rivers/streams (% of time)	89	94	98	100	98
TSS (<200 mg/l)	Water catchment streams (% of time)	99	99	99	100	99
	Non-water catchments rivers/streams (% of time)	100	100	99	100	100

Source: Annual Reports, National Environment Agency (NEA), Singapore.

The data of the past eight years shows that the inland waters in Singapore are conducive for aquatic life.

Handling hazardous substances

Under the Environmental Pollution Control Act of 1999, the Environmental Pollution Control (Hazardous Substances) Regulations and their subsequent amendments included as part of the Environmental Protection and Management (Hazardous Substances) Regulations, surprise audits of hazardous substances stored are conducted. Table 3.3 presents data from the past seven years and shows a tendency to reduce the use of verbal warnings.

The data also indicates that records have not been kept in order in less than 5 per cent of the cases.

Addressing complaints regarding water pollution control

The Pollution Control Department has a robust system to investigate the complaints received and, depending on the illegal discharges, punitive actions are taken against the polluters. Table 3.4 presents a number of complaints received over past eight years under various categories and the result of the enquiry. The figures shown in brackets refer to the number of substantiated cases.

The data shows that the actual number of incidents or complaints that have been corroborated has gone down over the years, whilst those related to water pollution (except for farm waste) has been increasing. The highest number of complaints relates to industrial wastewater followed by those regarding oil and/or chemical pollution. Nonetheless, a yearly average of only 16.5 water pollution incidents indicates an excellent enforcement record in a highly industrialized economy.

Further thoughts

Policies, institutions and legislations have constituted the tripod of water pollution control in Singapore. Their synergy has been the backbone of successful water

Table 3.3 Surprise inspections of hazardous substances

Year	No. of surprise inspections	Found to be not in order	Action taken		
			Legal action	Written warning	Verbal warning
2003	999	63	6	49	8
2004	1,036	37	1	34	2
2005	886	47	0	46	1
2006	999	58	2	55	1
2007	1,047	37	3	34	0
2008	822	41	6	35	0
2009	827	42	15	27	0
2010	888	58	8	50	0

Source: Annual Reports, National Environment Agency (NEA), Singapore.

Table 3.4 Disposal of water pollution complaints

Type of water pollution	No. of complaints in respective years and (no. of complaints or incidents that have been corroborated)								
	2002	2003	2004	2005	2006	2007	2008	2009	2010
Chemical/ oil	50	33	56	62	37	51	84	85	28
	(19)	(4)	(17)	(12)	(5)	(3)	(6)	(3)	(2)
Industrial wastewater	3	12	30	54	52	105	170	173	323
	(1)	(3)	(6)	(4)	(3)	(10)	(2)	(1)	(7)
Farm waste	2	0	0	0	0	7	3	1	0
	(1)	(0)	(0)	(0)	(0)	(2)	(5)	(3)	(0)
Domestic wastewater	22	16	2	8	20	14	25	29	8
	(7)	(2)	(0)	(1)	(1)	(2)	(0)	(0)	(0)
Others	34	30	7	7	7	2	102	103	9
	(7)	(11)	(3)	(0)	(0)	(1)	(3)	(2)	(0)
Total	111	91	95	131	116	179	384	391	368
	(35)	(20)	(26)	(17)	(9)	(18)	(16)	(9)	(9)

Source: Annual Reports, National Environment Agency (NEA), Singapore.

management in the city-state. The analysis shows the relevance, appropriateness and impact of the efforts made to develop and fine-tune laws and regulations as well as to structure and restructure several institutions in anticipation and response to Singapore's evolving economic, social and environmental needs.

Regulations passed in the Parliament and their effective implementation have led to the overall promotion of economic and social growth, the reduction of social inequalities and improvements in quality of life. All of this while consistently providing for the conducive conditions that make the country attractive to foreign investment but allows it to maintain its strong environmental requirements. In terms of water resources, the results show continuous improvements in water pollution control over the years and therefore the protection of this finite and valuable resource so important for the well-being of the city-state.

From the early 1960s on, and in spite of the country's urgent need for foreign investment to ensure industrial and commercial development, Singapore has made sure that environmental standards are not compromised. It has increasingly developed regulations, strengthened enforcement, instituted proper procedures for managing hazardous substances and ensured that industries incorporate adequate pollution control facilities when designing and constructing their facilities and that these are properly operated and maintained on a regular basis. Transparency, equal opportunities, infrastructural facilities and enforcement have ensured that, in spite of rigorous environmental measures, industries have complied with the legislation.

One less discussed aspect of success of effective enforcement of pollution control legislations in Singapore has been the economic benefits the city-state has received by adhering, and some time exceeding, international standards for water supply, sanitation and wastewater treatment. It is now universally acknowledged that investment in water supply and wastewater treatment reduces health and social costs significantly. Equally, there are significant costs associated with inaction in these areas (OECD, 2008).

Enforcement of legislation and provision of quality services does not happen without planning, institutional capacity and political will. Even in the late 1960s, the question of safe potable water supply and sanitation facilities was not about technical knowledge and capability of treatment processes but about willingness: on the part of authorities to plan and provide them, and on the part of the population, not to accept pollution as an inescapable consequence of urban and industrial development. Singapore from the beginning denied pollution as an inevitable offshoot of economic development.

Preventing water pollution has been interwoven with the city-state's urban growth through proper land use planning, control of building standards, development of environmental infrastructure and implementation of stringent legislation. Law enforcement has been assured by regular inspections conducted to guarantee proper and efficient operations and the maintenance of pollution control facilities; prompt actions in response to filed complaints; a competent legal system; and a corruption-free institutional setting. Additionally, regular monitoring of the waters has been carried out to supervise the adequacy and effectiveness of the different control programmes.

The institutions involved in water management, including the APU, PUB or the Pollution Control Department have all been involved in formulating policy principles, regulations, amendments to existing regulations, penalties and standards to be applied. Examples of this can be seen in the modifications made to various water-related legislations over a given period of time. They do not only demonstrate pragmatism but also serve to testify to the role these institutions play in influencing the policy-making processes and thereby directing the trajectory of implementation.

By ensuring timely, appropriate and adequate amendments in both laws and regulations, Singapore's water-related legal framework has remained relevant and effective to solve the country's problems spurring from industrial growth and urban sprawl. The analyzes show that the island's successful water management, and the body of appropriate and adequate legislation it has formulated and effectively enforced have been a crucial element for development.

Clearly, long-term planning has been essential. It has provided the framework for policies, regulations and institutions to operate jointly, achieving steady improvements throughout the years. One should not, of course, disregard the value of efficient interagency coordination which, even if not free of complexities and difficulties, has allowed the implementation of the different sets of public policies.

Singapore has seen the holistic development 'picture' and has shown the best of willingness and dispositions to invest in the required infrastructure, technological advancements and skilled personnel necessary for such an endeavour. Undoubtedly, whatever investments the city-state has made have not been in vain since they have brought about numerous benefits to the country and its people. Willingness to face the challenges of a complex developmental journey has stood as the difference between succeeding and failing. And even though every process can be improved in many ways, Singapore decided a long time ago that it had the will to acquire, build and develop the means required to embark on the journey towards sustainable development.

4 Managing water demands

Introduction

Water demand in Singapore has been increasing steadily along the years similarly to what has often been witnessed in most other countries. As population grows, standards of living improve, and the progress of urbanization brings about the acceleration of commercial and industrial developments, water demands have the tendency to rise unless specific policy countermeasures are taken. Moreover, socio-economic development expands the use of labour-saving equipment like washing machines, further driving up urban water use. As water demands escalate, the majority of countries have preferred the general approach to steadily increase water supply. Most unfortunately, managing water demands has not received similar political, social or technical attention.

Around the time of its independence in 1965, the situation in Singapore was very similar to what has been witnessed in other nations undergoing rapid growth. From a daily average consumption of about 32.5 Mgal/day in 1950, demand increased 2.52-fold to 81.9 Mgal/day by 1965, and to 110 Mgal/day by 1970. During these two decades, Singapore's policies focused on supply management to tackle its serious urban water management problem. For example, when the island faced a prolonged drought in 1960 and unusually low rainfall in 1961, thousands of Singaporeans woke up on 19 August 1961 to find there was not even a single drop of water available. The government reacted by instituting a strict water rationing system.

A small city-state like Singapore, with a land area of only 581 km^2 at the time of independence, has struggled to collect rainfall and find the space and favourable topography to construct large and deep reservoirs to store water and meet yearly needs. Even now, with a 23 per cent bigger land area of 714 km^2 brought about by land reclamation, the country continues to face serious land constraints to collect and store enough water to meet domestic needs. Therefore, with a limited array of options available to increase water supply, the island also began tackling demand.

This chapter analyzes Singapore's water demand management strategies from 1965 and how pricing, mandatory water conservation requirements and public education have been used as tools to respond to its water needs at different times in history (education and information strategies are analyzed in depth in Chapter 5).

Water demand management: historical background

Singapore is situated in the equatorial rain-belt and receives 2,358 mm of average annual rainfall. Had it been a country with a larger landmass, this level of rainfall would have been more than adequate to satisfy its water needs as long as they were properly managed.

Before the 1950s, the concept of Singapore as an independent country was never seriously considered. This small port formed the heart of the British Empire in South-east Asia. After the Second World War, the British detached Singapore from what was then called Malaya, now Malaysia, in anticipation of granting it independence in the foreseeable future. The plan was to keep Singapore as a crown colony so that the British naval bases and trade routes connecting Southampton, Gibraltar, Malta, Suez, Aden, Colombo and Hong Kong could be maintained. However, the loss of the Suez Canal in 1956 changed previous plans. An alternative strategy was considered to join Singapore with the Federation of Malaysia and the British territories of North Borneo (now Sabah) and Sarawak into Malaysia. However, less than two years after the merger, Singapore was asked to leave the newly formed country and was on its own. This separation and the events that led to it created many tensions between the two countries.

Once outside the Federation, Singapore had to suddenly figure out the terms of its survival as an independent country as plans had always been made thinking of the island as part of a much larger entity. This separation created all types of complex and difficult management problems as Singapore had basically become a city-state without any natural resources. Water supply was one of the most difficult issues that had to be addressed.

Foreseeing such difficulties and with a clear vision in mind, Singapore's leaders insisted that the separation agreement included a guarantee that Malaysia would honour the two water accords to import water from Johor. Given its foremost importance, this issue could not be treated as a separate matter between the two counties, and it was thus explicitly mentioned in the Separation Agreement signed between the governments of Singapore and Malaysia on 9 August 1965. Such document guaranteed the 1961 and 1962 water agreements were going to be respected (for more information on this issue, see Chapter 7).

The seriousness of the water availability situation can be realized by the fact that in 1965, when Singapore became independent, it had only three reservoirs (MacRitchie, Peirce and Seletar). These internal sources were small by international standards and could be used to draw less than 20 per cent of the island's water needs. Unsurprisingly, meeting as much of the country's water demand from purely national sources became paramount and a national strategic priority.

Already at the time of the Japanese invasion during the Second World War, Singapore had survived a major water shock and witnessed the extent of its water vulnerability when the pipe supplying water from Johor over the Straits of Johor was blown up. After independence, the issue of foreign reliance became once again a matter of forefront importance as Malaysia's leaders were acutely aware of Singapore's unhealthy dependence on water imports from Johor. On the day of Singapore's independence, Malaysian Prime Minister Tunku told the British High Commissioner to Malaysia, Anthony Head, that unless Singapore listened to Malaysia and did as suggested, Malaysia would cut off the water supply. Head, as Tunku had correctly surmised, promptly passed this warning to Lee Kuan Yew (personal interview with Lee Kuan Yew, 11–12 February 2009).

This threat can be corroborated by a confidential telegram sent by the Secretary of State for Commonwealth Relations, Arthur Bottomley, to the British Foreign High Commissioner in Canberra, transmitting the original telegraph Anthony Head had shared with Singapore's political leadership:

> I saw Tunku at 9 a.m. this morning and gave him Prime Minister's telegram. He said that it was too late to change now and that he was just off to tell meeting of Alliance Members of Parliament. He was sorry that he had not informed us before but he was confident that it was no good arguing about step, which in his view was inevitable. He could see no other course to take. He would reply to Mr. Wilson's message as soon as possible.
>
> I said that I thought that in middle of confrontation[1] and with our deep involvement in Malaysia and its future it was most surprisingly that we had not even been given time to express view or to discuss full implications of so drastic a step. I said for instance that I presumed that now Lee would have full autonomy in foreign policy. Tunku confirmed this. I said that one could easily envisage possibility of Singapore Government pursuing foreign policy, which might put us in most embarrassing position. For instance what would happen if they decided to disassociate themselves from confrontation. Tunku said that if Singapore's foreign policy was prejudicial to Malaysia's interests they could always bring pressure to bear on them by threatening to turn off the water in Johor. With this startling proposal of how to co-ordinate foreign policy we turned to question of Borneo.
>
> (National Archives of Australia, 1965)

This fact is also mentioned on a confidential letter from the UK Southern and South East Arts (S&SEA) Research Department to the South-west Pacific Department on 1 June 1971 (Feirn, 1971).

The threat was a wake-up call for the political leaders and people of Singapore. Prime Minister Lee Kuan Yew immediately called Lee Ek Tieng, engineer in the City Council and Public Works Department at that time, and asked him to estimate Singapore's annual rainfall, the technical feasibility to capture every drop, and whether this measure could make the island self-sufficient in water (personal interview with Lee Kuan Yew, 11–12 February 2009). This was the beginning of a national strategy to attain water security by catching and storing as much water as possible nationally. Concurrently, measures also included reducing usage, including per capita domestic consumption as much as possible and dramatically bringing down the level of unaccountable losses to the minimum possible and as soon as it became feasible to do so.

While the Public Utilities Board (PUB) had expanded the catchment and storage capacity of Seletar Reservoir (finalized in 1969) and completed the Upper Peirce Reservoir scheme in 1975, it was not possible to expand protected catchments (those that are left in their natural states as far as possible and development is not allowed) indefinitely. To obtain as much water as possible from national sources, the smaller streams and rivers were dammed up first, with big and highly polluted rivers left for later actions. Singapore had two motivations to clean up its big rivers. First was the need to obtain as much water as possible from national sources. Second was that during British rule, areas inhabited by the colonial population were clean and beautiful but the remainder of the island was somewhat squalid, often with open sewers, squatters and very limited municipal facilities. With a 'one man, one vote' system, Prime Minister Lee decided that, if the party wanted to avoid losing votes and elections, uniformly good environmental conditions had to prevail for the whole island and for all of its people. Accordingly, the idea of a 'clean and green Singapore' was born, and it became an integral component of the national strategy and the efforts for water self-sufficiency.

Lee Kuan Yew became the only Prime Minister anywhere in the world in recent history taking special and continuing interest in water throughout the entire 31 years he was in office. During this period, he personally and regularly received all relevant water news, and the water situation was coordinated directly from his office. No ministry could make any decision that could in any way jeopardize the country's quest for water security: they were simply vetoed by the Prime Minister (personal interview with Lee Kuan Yew, 11–12 February 2009).

Because of such consistent support from the highest political level, by the time Prime Minister Lee retired from office, the country had managed to put in place one of the most efficient and effective water and wastewater management systems in the world. It had the main objective to increase water supply by every means available, and whenever and wherever possible. By managing demand, Singapore ensured that the country made spectacular advances towards water security, in spite of curtailing demand being politically more difficult to attain than expanding water supply.

Evolution of water demand management

Like in all countries, the concept of water demand management and its application has been an evolutionary process in Singapore. During the 1960s and 1970s, as almost everywhere, water demand management was not seriously considered as a policy option. On the contrary, steady increases in water demand were considered to be a good indicator of economic growth and national development. For example, in 1965, the PUB emphasized its contributions to the overall economic development of a newly independent Singapore since it was able to meet the increasing water demands resulting from accelerated rates of urban expansion and housing construction (PUB, 1965). The same line of thought was also evident in the PUB's 1967 claims that the processing and sale of potable water had been steadily increased to satisfactory levels where demands had been covered in spite of reductions in the rate of water demand increase during the previous five years (PUB, 1967). By 1969, higher production and sales of water, electricity and gas were considered to be a positive sign of progress because these increases were considered as indicators of rapid industrial, commercial and housing developments, consistent with rising standards of living (PUB, 1969).

Following a serious drought in 1971, the concept of intensified water demand as a sign of progress started to change. A water saving campaign was started that same year, and exhibitions were mounted for the following three on the theme 'Water is Precious'. Moreover, in 1973, water tariffs were modified and an increasing block tariff system was introduced to raise the PUB's total income and prevent wastages in the domestic sector. However, even at this point, the main rationale behind setting higher tariffs was not water conservation. The official explanation offered to the public was that Singapore received most of its water supply from Malaysia and this dependence had to be reduced. The chief motivation behind adjustments in water tariffs was the need Singapore had to invest more and more funds to develop expensive storage, treatment and distribution capacities (PUB, 1973). The extent to which higher water tariffs could reduce water demands was mostly a secondary objective.

Due to the continuous increase in water consumption, the Water Conservation Plan was introduced in 1981. It had the objective to reduce projected water demand for the island. This plan represented a radical departure from previous thinking regarding water policy. In contrast to the 1960s and 1970s when higher water demand was seen as a sign of economic development and an indicator of higher standards of living, from 1981 onwards the PUB's Annual Reports extolled the virtues and importance of water conservation. Analysis of the Hansard records on the discussions in the Singapore Parliament further indicates the growing realization of the importance of water conservation measures to progressively reduce future water demands.

Along the years, heightened emphasis on water demand management was mostly a response to the cost of additional water supplies. In the 1960s, there were abundant water supply options to explore and develop. Even though there was a limit to the number of projects that could be developed, it was a relatively straightforward solution to expand water supply as fast as possible and raise water prices to ensure sufficient revenue to cover the costs of these developments. Nevertheless, by the 1980s, Singapore was already building what was considered to be the last reservoir scheme at Sungei Seletar-Bedok. The next water resources were projected to come from desalination, an alternative that was ten times more expensive than using rivers' surface water at that time. It was then that the need for water conservation struck a chord with leaders, planners and water professionals.

Water pricing

The increased focus on water demand management that began in the 1980s can be seen in terms of both pricing for cost-recovery and non-pricing measures. The

Singapore government had repeatedly stated that water prices had to be increased to cover the cost of growing water demand, with the objective of higher tariffs being mostly for cost recovery. For instance, during the 1980s and 1990s, the cost of water supply and wastewater services represented less than 0.5 per cent of average household income, which meant that water pricing was not an incentive to reduce domestic water demand. The water tariffs during the 1980s and 1990s were decided primarily to ensure cost recovery.

Singapore, similar to other British colonies, had inherited the British water system of charging for water. As such, the challenge was not so much to charge for water but rather to get the population to accept increases in tariffs to assure cost recovery, as well as restructuring the tariffs so that the non-domestic sector would no longer subsidize the domestic sector. The latter objective was achieved in 1997.

In 1991, an explicit tax aimed at conserving water was put in place. While this marked a milestone in the use of water pricing to reduce water consumption, the basis for determining the tax quantum was not clear. In 1997, Singapore explicitly started to use economically efficient price signals to manage water demand. The consequent pricing revision was aimed at recovering the full cost of production and supply through the water tariff as well as to reflect the higher cost of alternative water supply sources through the water conservation tax. The revenues generated from this particular tax were not allocated to the PUB because the agency would be receiving funds it did not need for water production and supply. Instead, they were transferred to the consolidated fund of the government, which was managed by the Ministry of Finance (Tan *et al.*, 2009).

While the water conservation tax has been levied to encourage water conservation, some other non-water related taxes have also been levied in terms of water price. One example is a statutory board tax that was introduced in 1969 to defray the increasing defence expenditure that was required to build up the Singaporean armed forces in response to the British announcement that all troops would be withdrawn from the island by 1971. This 10 per cent tax was levied from the total utility bill, which at that time comprised electricity, gas and water. From a policy point of view, this tax had no relation with water even though it may have had an impact on the consumers' behaviour since it affected their water bill.

Changing water tariffs, 1965–2012

Water price (water tariff and related taxes) were revised more than ten times between 1965 and 2012. As noted earlier, the initial price reviews were primarily motivated by the need to recover costs, rather than to encourage water conservation.

During the 1960s, consumers were charged a flat volumetric rate with a fixed charge on meter rents and a turn-on fee. Non-domestic consumers were divided into three categories: shipping, consumers that processed water for sale, and others. In 1965, it became clear that the fuel oil tax increased fiscal contributions more considerably than property taxes and that the large capital investments required to meet increasing electricity, water and gas demands would necessitate a tariff revision for all services. Thus the first tariff revision since 1954 came into effect in November 1966 (PUB, 1966; *The Straits Times*, 22 October 1966). As a result, the price of water per thousand gallons for domestic consumption increased from 60 cents to 80 cents, water supplied to ships increased from $3.75 to $4.00, and that to industries processing water for sale was revised from $2.00 to $2.50. For other trades and industries, the price increased from $1.30 to $1.50 per thousand gallons. For the government, the price was raised to $1.00 and for statutory boards and the foreign armed forces, it became $1.50 (PUB, 1966). The reasons behind the tariff raise were explained to the public over the mass media, using TV, radio and newspapers, and were generally accepted by all consumers (PUB, 1966).

Rates were revised so that increased costs for water services were spread across all consumers and to avoid imposing a burden on the industries that underpinned the economic development of the country (PUB, 1966). An additional reason was to ensure that, as far as possible, electricity, water and gas supplies were self-sufficient with their own generated revenues. When reporting the impacts of the 1966 tariff revision, the PUB emphasized that this measure had led to an increase in revenue, mentioning only in passing the impacts it had on managing water demands. This once again reinforced the notion that price adjustments served almost exclusively cost-recovery purposes and were not directly related to water conservation (PUB, 1967).

Parliamentary budget discussions in December 1967 on water pricing noted that this had remained static since 1954 and until 1966 while the cost of production had increased continuously. In 1967, these expenses were expected to augment more to $0.7842 per thousand gallons from $0.7294 in 1962. As a result, it was necessary for the PUB to increase its water tariffs as it was up to the Board to guarantee repaying the loans obtained from the World Bank and the Singapore government to finance public utility projects (Hansard, 1967).

In 1969, it was planned to extend sanitation services to the entire population in the city-state. The cost of the sewerage projects would be approximately $60 million for the 1968–1973 period for which the World Bank would provide a loan of $18 million to cover the scheme's foreign exchange component (Goh, 1970a).[2]

Since this loan was conditional upon the revision of sewage fees, water tariffs were increased from March 1970: domestic users would pay an additional 20 cents per thousand gallons of metered water in addition to a $2 monthly fee per sanitary fitting; non-domestic premises would be charged 50 cents per thousand gallons of metered water; and, in the case of industries using water in their end-products, meters would be fixed to assess the volume used for industrial purposes and not discharged into sewers (Hansard, 1969a, 1970; PUB, 1970).

It was then decided that the PUB would collect the fees on behalf of the government. The additional payments from domestic premises would vary between 20 cents and $1 per month for more than 55 per cent of the population, while another 30 per cent would pay between $1 and $2 per month. Commercial and industrial premises would cover a percentage of the costs, varying from $2.50 to $15 per month in the case of coffee shops, and $15 to $75 per month for smaller industries. For the increases in utility charges not be absorbed by poorer households, the prevailing PUB sales tax exemption on consumers with monthly bills below $10 would be extended to those with a consumption below $12 a month. Similarly, the 5 per cent tax payable on bills between $11 and $20 would be applicable to those between $12 and $25 a month and the 10 per cent tax rate would only apply to bills over $25. In spite of these increases, the return on investments in sewerage was expected to be no more than 2 per cent on the capital invested (Hansard, 1969a).

All registered water consumers, including those who did not have access to sanitation and who were then mostly located in rural areas, would have to pay the additional charge. It was assumed these dwellers would be less reluctant to abandon the bucket-system if they had to pay a charge, even if small, for a sanitary service they did not have. In order to be fair to those consumers, the Public Works Department would have the duty to extend sewerage facilities to all consumers as soon as possible (Hansard, 1970b).

In 1973, Singapore modified the domestic water tariff structure from a flat volumetric rate to an increasing block tariff for the very first time. There was an escalating cost for every block of 25 m^3 to 75 m^3 with the aim of reducing water wastages in the domestic sector. Even when it was argued water conservation was the reason for changing the rate structure, cost recovery continued to be the main driver. That same year, and for the first time, non-domestic consumers were added as a specific category. The government also introduced three sets of tariffs for different household sizes living in one block unit so that large households with a higher consumption rate would not fall into higher tariff blocks. Block tariffs were established after analyzing the household average per capita water consumption so that a minimum necessary amount of water could be provided at low rates (PUB, 1973).

Both the domestic and non-domestic water tariffs were revised again in 1975 due to the significant increase in water consumption and the accompanying growing capital and operating costs, as well as to encourage economic development (PUB, 1975). Tariff-generated revenue was necessary to pay for the Upper Peirce Reservoir and Kranji-Pandan Reservoir scheme, two major water supply projects that had an approximate total cost of $137.3 million. Higher fees were also to cover the summing costs of treating raw water from the highly polluted Kranji-Pandan Dam (Hansard, 1976).

In 1981, several measures were introduced to encourage a more rational water use. The three sets of domestic water tariffs introduced in 1973 were eventually simplified and replaced by a single domestic tariff in 1981. Under the then regulations and at its discretion, the PUB could grant concessions to any premises where there were two or more households and where more than ten persons lived in the same place. The then four-tier domestic tariff was revised to three tiers. The non-domestic water tariff structure was also revised in 1981 from a flat volumetric rate to an increasing block rate where consumers with a water consumption exceeding 5,000 m^3/month had to pay a higher rate (PUB, 1981).

Previous to these rises in the water tariff, Goh Chok Tong, then Senior Minister of State for Finance, referred to the intensification of water consumption in a Parliament session (Goh, 1981). Goh noted that water demand had grown very rapidly, by 7.4 per cent in 1979 and 7.7 per cent in 1980. An important concern was that if the consumption of water continued to increase at such high rates, shortages would be faced in times of crisis (including droughts), given the limited water storage facilities in the city-state. The Minister also explained that such rapid increases in water use could be reduced by either water rationing or through price mechanisms, noting how the latter was 'the lesser of two evils' (Goh, 1981: column 248). This was exemplified by the 1973 and 1975 water rate revisions by which water demand had been immediately reduced. On those two occasions, water consumption increases dropped drastically from 7.7 per cent in 1972 to 1.6 per cent in 1973 and 1.8 per cent in 1974. In 1975, water consumption had increased by 9 per cent but then decreased to 1.7 per cent in 1976 after the price was revised. The Minister concluded that price revision 'had a salutary effect on wastage of water' (Goh, 1981: column 248). In order not to adversely affect those households belonging to the lower income group, their minimum amount of water required was taken into consideration in any water rates revision.

In 1983, the three separate categories of water users (government, statutory boards and armed forces) disappeared and they were charged the non-domestic rate. Three new categories were introduced, namely hotels, restaurants and construction sites, and were charged rates equivalent to the second tier of the non-domestic water tariff (flat rates of 100 cents/m^3). The water tariff was increased again to cover the costs of constructing new reservoirs and treatment plants. For domestic consumers, it increased from 7 to 20 cents/m^3 depending on their consumption and for commercial and industrial consumers, it increased from 20 to 40 cents/m^3 depending on the nature of their activities (PUB, 1983). By 1986, there were four consumer categories namely domestic, non-domestic, shipping and water processed for sale. The non-domestic tariff structure was simplified, reverting to a flat volumetric rate pegged to the highest domestic tariff tier. According to the PUB, water rates were raised to meet the mounting cost of producing and supplying water (PUB, 1986).

Subsequently, and as mentioned before, the water conservation tax was introduced in 1991 as a potential pricing tool to discourage excessive water consumption (Tan *et al.*, 2009). Thereafter, water price has included both the tariff as well as the conservation tax. The tax was initially applied to domestic consumption above 20 m^3 and to non-domestic consumption from the very first drop. In 1992, concerns regarding the development of strategies to reduce increasing water consumption were again expressed in the Parliament (Hansard, 1992). The need to promote higher GDP without necessarily using more water was discussed, especially as the resource was still mostly imported from Malaysia. The objective was that the economy could grow without concomitant higher water use. By then, and as part of this strategy, the water conservation tax had already been in place from the year before but even then water demand continued to soar. In fact, in

1991, domestic demand increased by 5.8 per cent, which was much higher than the annual average 3.8 per cent increase of the previous five years. The Parliament once again considered the desirability of elevating the water conservation tax even further.

Perhaps the most significant change in terms of water pricing policy occurred in 1997. This was the first time the government attempted to price water based on economic efficiency and aiming, in the long term, to restructure both the water tariffs and the conservation tax into uniform flat rates. As much as possible, the price per cubic metre of water was to be the same irrespective of the user (household, industry or a construction site), and regardless of the amount consumed (PUB, 1997a). Thus, several significant new changes were introduced. First, water price was pegged to the cost of desalinated water to reflect the higher cost of alternative supply sources. With this, Singapore became one of the pioneering countries introducing marginal cost pricing. Second, the water conservation tax was charged from the very first drop delivered, even for the domestic sector, to signal the importance of water as a strategic resource for the country (PUB, 1997b). Third, the lowest two domestic consumption tiers (0–20 m^3 and 20–40 m^3) were merged into one, domestic consumption above 40 m^3 was charged a higher tariff and the non-domestic water price was made equivalent to the lower domestic consumption tier rate. Finally, the volumetric sewerage fees were also increased to link the fee charged with the volume of wastewater generated. To avoid drastic changes and confusion among consumers, this new pricing structure was gradually implemented over a period of four years, from 1997 to 2000.

Since then, there have been no price increases, although in 2007, the government mentioned plans to introduce a single price for water supply and distribution and for wastewater collection, treatment and disposal. The government also announced the intention to remove the fixed sanitary appliance fee and thereby move to a fully volumetric charging mechanism for the waterborne fee that would be more equitable.[3] The thinking was that the cost of used water treatment should be pegged to the volume of wastewater generated and not to the number of sanitation facilities installed.

Singapore has introduced rebate schemes throughout the years to make sure that the impact of higher living costs on lower income households is minimized. To this end, a Citizens' Consultative Committee (CCC) assistance scheme was established in 1994 following the introduction of the Goods & Service Tax (GST). The aim of this assistance scheme was to support lower income households who, despite a comprehensive offset-package (including reductions in personal income tax and property tax), were still unable to pay GST. Similarly, and following revisions of electricity tariffs in 1996, rebates were introduced on public utilities bills to assist lower income households (Ministry of Finance, 1996). These rebates were placed in qualifying households' utilities account, and could be used to pay utility bills, including water. If the rebates were not used completely in the first month, they could be used in subsequent months. The government chose to decouple assistance from consumption, so that it would not lead to over-consumption of water.

Moreover, utility rebates were maintained following the 1997 revisions of the electricity tariff and adjustments to water tariffs and the water conservation tax. The CCC assistance scheme was also extended to those households affected by higher utility bills but unable to benefit from rebates (Ministry of Finance, 1997). Tables 4.1 and 4.2 summarize the different and aforementioned water tariffs and average cost per cubic metre from 1965 to 2000 as well as the water conservation tax rates from 1991 to 2000, respectively.

The change in domestic consumption over time, indicating the years when a price increase has been introduced, is shown in Figure 4.1. The figure also shows the domestic consumption 20 m³ water bills, daily per capita water consumption and responses to each price increase and to the different public education efforts that have been implemented.

Pricing, mandatory and technical measures as well as ongoing education out-reach programmes have all been part of a water conservation message. These efforts have resulted in reductions in domestic per capita consumption of 19 litres/day, decreasing from 172 litres/day in 1995 to 153 litres/day in 2011. In addition, industries have been encouraged to become more efficient and substitute potable water with NEWater. Conservation actions have also made Singapore one of the cities in the world with the lowest amount of unaccounted-for water (below 5 per cent). Even with all of these positive developments and achievements, however, water conservation measures need to be given more thorough consideration for

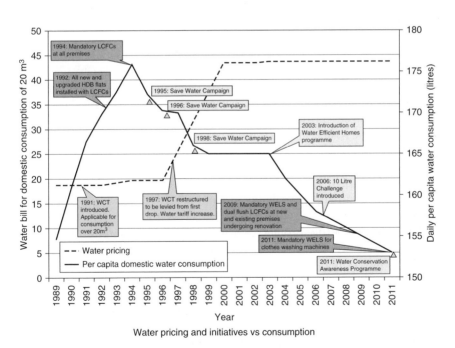

Figure 4.1 Impact price changes and initiatives have had on per capita domestic water consumption from 1989 to 2011.

Table 4.1 Water tariffs and average cost per cubic metre from 1965 to 2000

Year	1965	1966	1973	1975	1981	1983	1986	1993	1997	1998	1999	2000
Description	(From 1 January)	(From 1 November)	(From 1 February)	(From 1 September)	(From 1 September)	(From 1 December)	(From 1 August)	(From 1 December)	(From 1 July)	(From 1 July)	(From 1 July)	(From 1 July)
Domestic												
Consumption block (cu m per month)												
1st block			1–25@ 0.22	1–25@0.30	1–20@0.35	1–20@0.24	1–20@ 0.53	1–20@0.56	1–20 @0.73	1–20 @0.87	1–20 @1.03	1–40@ 1.17
2nd block			25–50@ 0.26	25–50@0.40	20–40@0.45	20–40@0.57	20–40 @0.75	20–40@ 0.80	20–40 @0.90	20–40 @0.98	20–40 @1.06	
3rd block			50–75@ 0.33	50–75@0.50								
Follow-on rate	0.13	0.18	Above 75 @0.44	Above 75 0.66	Above 40@ 0.75	Above 40 @0.95	Above 40@1.10	Above 40 @1.17	Above 40@ 1.21	Above 40@ 1.24	Above 40@ 1.33	Above 40@ 1.40
Concessional rate for two or more households + 10 < Number of persons < 20												
Consumption block (cum per month)												
1st block			1–50@ 0.22	1–50@ 0.30								
2nd block			50–100@ 0.26	50–100@ 0.40								
3rd block			100–150@ 0.33	100–150@ 0.50								
Follow-on rate			Above 150@0.44	Above 150 @0.66								
Concessional rate for two or more households + 20 < number			1–75@ 0.22	1–75@0.30								

of persons < 30												
Consumption block (cu m per month)												
1st block			75–150 @0.26	75–150@ 0.40								
2nd block			150–225 @0.33 Above 225@0.44	150–225@ 0.50 Above 225 @0.66								
3rd block												
Follow-on rate												
Non-domestic	0.29	0.33	0.44	0.66	1–5,000@ 0.75 Above 5,000@0.85	1–5,000@ 0.95 Above 5,000@1.10	1.10	1.17	1.17	1.17	1.17	1.17
Others												
Shipping	0.82	0.88	0.88	1.32	1.55	1.95	1.95	2.07	1.99	1.99	1.92	1.92
Water processed for sale	0.44	0.55	0.88	1.32	1.55	1.95	1.95					
Government		0.22	0.44	0.66								
Statutory boards		0.33	0.44	0.66								
Armed forces		0.33	0.44	0.66								
Hotels					0.85	1.10						
Restaurants					0.85	1.10						
Construction sites					0.85	1.10						

Table 4.2 Water conservation tax from 1991 to 2000 (rate based on percentage of water tariff)

Year	1991	1992	1995	1997	1998	1999	2000
Description	(From 1 April)	(From 1 April)	(From 1 April)	(From 1 July)	(From 1 July)	(From 1 July)	(From 1 July)
Domestic							
Consumption block (cu m per month)							
Below 20	—	—	—	10	20	25	30
20–40	5	10	15	20	25	30	30
Above 40	5	10	15	25	35	40	45
Non-domestic	10	15	20	25	25	30	30
Shipping	10	15	20	25	25	30	30

long-term sustainability as the country still imports more than 50 per cent of its water resources (Balakrishnan, 2012).

Mandatory measures

As early as the 1940s, the Water Department part of the City Council had a Waste Detection Section that was responsible for identifying leaks in the water supply network. The acknowledged importance of minimizing leakages led the Waste Detection Section to conduct house-to-house inspections, quantify metered and unmetered leakages, and estimate savings for leakages detected. From 1978 onwards, engineering solutions were applied at the consumers' end to reduce water demands. That year, the PUB embarked on a pilot study to test how much water could be saved by installing thimbles (PUB, 1978).[4] This document anecdotally showed that savings in water demand of about 10 per cent could be realized but that more studies were required. Consequently, the installation of thimbles continued from that year onwards. Forthcoming PUB studies found such technology could achieve water savings as high as 11.6 per cent (PUB, 1980).

As a result of ever rising water consumption, the Water Conservation Plan was drawn up in 1981 and a Water Conservation Unit established to implement the initiatives detailed in the plan. The plan focused mainly on large consumers, on those sectors where the scope for saving was considered to be the greatest and on the top 4 per cent of the domestic sector (PUB, 1981). That is, the main focus for the 1980s was the non-domestic sector. Unlike the Waste Detection Section that carried out engineering checks on leakages and wastage, the Water Conservation Unit had the mandate to focus on actions reducing water demand from the con-sumers' end. These initiatives included technical measures, both mandatory and voluntary, and public education. Water audits and economic incentives chiefly targeted the non-domestic sector.

In 1983, the PUB made major changes to its water supply regulations to reduce non-domestic water wastage in public places and to improve plumbing standards (PUB, 1983). With these amendments, it became mandatory for non-domes-tic consumers and common areas of all private high-rise residential apartments and condominiums to install water-saving devices in their premises such as self-closing delayed action taps and constant flow regulators. This measure was aimed at reducing excessive flow rates at the taps and preventing water waste by users who did not pay directly for the water. At the same time, a maximum flow rate for water fixtures such as taps and overheads was imposed on all non-domestic premises (maximum flow rate for domestic premises was imposed until 2002).

Since 1992, low capacity flushing cisterns (LCFCs) employing less than 4.5 litres per flush have been installed in all new public housing units. From 1997, using these devices over the conventional 9-litres cistern has become mandatory for all new and ongoing building projects, including all residential premises, hotels, commercial buildings and industrial establishments. In the 1980s, the Housing & Development Board (HDB) realized that the largest percentage of household water use, around 20 per cent, accrued to toilet flushing. Accordingly, there was a search for ways to find water-saving mechanisms. At that time, Sweden had already been using LCFCs that used 3–4 litres of water per toilet flush. Using cisterns that could to do the same job with less water thereby seemed like a natural solution. However, there was a concern that lower water volumes might slow down or even halt the flow of wastewater in the sewerage system, thereby creating public health problems. Reservations that the lower flow rates might not work were also expressed. Questions were brought up regarding the efficiency of half discharges as compared to the full amount, suggesting water savings might not even be realized if users needed to flush more than once to effectively remove waste.

At that same time, a Gustavsberg local agent, a LCFC pioneer Swedish company, approached officers at the Ministry of the Environment (ENV), HDB and PUB to give them a demonstration on the new cisterns. The company, working through the Swedish Embassy in Singapore, subsequently invited officers from these three government bodies to visit Sweden to assess the efficiency of LCFCs and their acceptability among the local population. Even after the fact-finding visit, the ENV, HDB and PUB officers decided to carry out a nine-month study in actual HDB flats to record water savings and study LCFC performance under Singaporean conditions in terms of flush efficiency, visual attractiveness, possible defects and malfunction, impact of the reduced flow on the sewerage system and also the extent of water savings. Three Swedish LCFCs were used in the pilot study, which turned out to be a success. Positive results showed there was no problem with waste transportation in sewer lines and that there have been actual water savings of approximately 19 litres per person per day. Based on the HDB's calculations, if LCFCs replaced conventional 9-litre cisterns in all HDB flats, annual water savings could be equivalent to the storage capacity of Seletar Reservoir (more than 5,000 Mgal). Even though planners were convinced that the mandatory installation of LCFCs would have a significant impact on water savings, they continued implementing technical measures.

In the meantime, the HDB was going one step further to ensure that the market was ready for such a change. The three LCFCs used for the pilot study were Swedish and planners were thereby concerned that local cistern manufacturers would be negatively impacted. The HDB gathered feedback from seven organizations involved in the procurement and installation of sanitary devices (Singapore Institute of Architects, Singapore Sanitary Ware Importers and Exporters Association, Singapore Sanitary and Plumbing Association, Real Estate Developers Association of Singapore, Singapore Contractors Association Ltd, Singapore Hotels Association and Association of Consulting Engineers Singapore) to ensure the private sector supported the anticipated measure. These propositions were well received and led to some important adjustments that were key in the successful implementation of water-saving mandatory measures. For example, the Singapore Sanitary Ware Importers and Exporters Association voiced concerns that companies would need time to clear existing stocks of conventional cisterns and to acquire cisterns that would meet the new requirements. Thus, it was agreed that there should be a period between the announcement and cut-off date for the market to adjust to the new provisions. Finally, in 1992, LCFCs that used not more than 4.5 litres of water per flush were installed in all new HDB flats and upgraded units. By 1997, all new premises and ongoing building projects, including all residential premises, hotels, commercial buildings and industrial establishments, were also required to install the water-saving LCFCs. Twelve years later, in 2009, the PUB took a further step and mandated the use of dual flush LCFCs that use three litres for a reduced flush or 4.5 litres for a full one. The thinking was that dual flush LCFCs would use even less water as users were given a choice of using a reduced or full flush (PUB, 2008a).[5]

In addition to these mandatory measures, the voluntary adoption of water-saving devices was also encouraged. During the 1995 Save Water Campaign, free thimbles were distributed to reduce excessive pressure and flow by fitting them into household tap nozzles, showerheads and hose connections. Since 1999, thimbles have been installed in all washbasin taps at every single new HDB flat. In 2003, the maximum flow rates imposed on the non-domestic sector were reviewed and reduced between 25 and 33 per cent. These maximum flow rates were also imposed on the domestic sector for the first time.

Further measures have been put in place ever since. In 2006, the PUB launched the 10-litre challenge as an 'umbrella programme' seeking to get the population to reduce its daily per capita water consumption by 10 litres. As part of the initiative, the PUB and the Singapore Environmental Council introduced the voluntary Water Efficiency Labelling Scheme (WELS) to provide water consumption and efficiency information to help consumers make informed choices on their purchases of water fittings and products (PUB, 2008b). The Mandatory WELS, implemented in 2009, is a grading system of 0/1/2/3 ticks to reflect the water efficiency level of a product which applies to taps, mixers, dual-flush low capacity flushing cisterns, urinal flush valves and waterless urinals.

In terms of regulations, the PUB uses the Public Utilities (PU) Act (2002) and its subsidiary legislation in order to impose the mandatory water demand management measures (regulatory instruments are analyzed in depth in Chapter 3).[6] These provisions generally include using price and technical mechanisms to discourage water wastage, as well as punitive measures for non-compliance such as fines and court prosecution. For example, Section 23 (2) of the PU Act gives the PUB the power to discontinue water supply to a person should the Board find it responsible for misusing or wasting water. Moreover, Section 34 (1) states that anyone occupying a premise who fails to immediately notify the PUB of any pipe or apparatus obstruction or damage to the water supply network in or upon the establishment shall be liable, upon conviction, to a fine of up to $5,000. Severe penalties are also stipulated for anybody who wastes water under Section 50 (1), setting fines of up to $50,000 or imprisonment for up to three years, or both, and in the case of a continuing offence of water wastage, further $1,000 fines can be charged for every day or part thereof during which the offence continues after conviction.

Having the power to enforce compulsory measures is not the end of the story. If consumers were not aware of such provisions, it would be difficult for them to comply, and by extension, possibly unfair if they were prosecuted for acting based on lack of or insufficient knowledge. In general, beliefs and ideas about the long-term need to conserve water is a significant determinant of actual daily water use. Therefore, voluntary measures that include public education and outreach are amongst the most important initiatives. This aspect is discussed in detail in Chapter 5.

Further thoughts

The story of Singapore has been one of carefully and calculated long-term development planning. Contrary to what may have been thought in 1965, the apparent vulnerability to which the city-state is exposed to due to its small size, almost total lack of natural resources and high dependence on external water sources, has been turned into an opportunity for the island and its population to manage its resources significantly more efficiently than most other countries.

Long-term planning, pragmatic policies and their thorough implementation have allowed the city-state to find the right balance between national interests and their economic efficiency. Water has come be considered as a strategic element for survival first, and subsequently a factor for growth, development and improvement in quality of life. Assessing Singapore's journey towards sustainable development through water lenses leads to the conclusion that the pursued holistic management of water resources has enabled the city-state to progress towards its overall development goals.

Along the years, Singapore's water management has been challenged by geographical, developmental and environmental conditions, namely land restrictions for rainwater collection and sites for storage reservoirs; droughts, floods and pollution; and increasing per capita domestic and non-domestic use. The city-state has regularly looked for possibilities beyond the obvious, implementing short-, medium- and long-term plans, and searching for conventional and unconventional measures that promise to more and increasingly efficient opportunities to manage water.

As such, water supply and demand strategies have been tailored with time.

In the 1960s, the PUB's approach was supply oriented, instead of focusing on managing demand. Nevertheless, after the 1971 drought, and mostly during the 1980s, water demand management started to play an increasingly important role. Later, in the 1990s, the PUB started making use of economic instruments and community-based approaches to manage water demand.

It is generally regarded that since human behaviour is difficult to change, economic incentives can foster attitudinal shifts. For instance, setting and enforcing higher prices can encourage lower water consumption. This means that pricing could be further increased to reduce per capita consumption of water to much less than 153 litres. The current consumption rate in Singapore is still considerably higher in comparison to several European cities, some of which have already brought down their water consumption to near 100 l/capita/day (for example, Barcelona, Copenhagen, Hamburg).

Tariffs were last revised in 2000. Since then, water pricing not only has not changed but inflation has effectively lowered the real price of water. In contrast, the average household income has gone up, meaning the average proportion of water as part of expenditure has become smaller over the years. Most notably, for a country that imports more than half of its water from another country, from a water security viewpoint and due to future uncertainties such as those related to climate change for example, per capita consumption and industrial water demands are yet to be significantly reduced. As it has been proved that economic instruments can reshape consumption patterns and human behaviour, they should be given priority consideration in the future to further bring down water consumption.

5 Education and information strategies for water conservation

Introduction

Engaging the public and producing behavioural change is a fundamental part of any process seeking sustainability. With regards to water conservation, public engagement begins by providing information and creating awareness on the real need to reduce water use. Once this initial level of engagement has been reached, campaigns then try to create an interest in reducing water consumption with the aim to persuade informed and well-educated citizens to participate in or take action to conserve water at a further advanced involvement level.

The delicate interaction between social, economic and institutional frameworks, governance issues and water systems (including physical, chemical and biological factors) defines the degree of interactive management among the different parties and interests. In the following discussions, this degree of interaction will be referred to as 'level of public engagement' where communication and interface heightens depending upon the governance style and the roles stakeholders play within a given socio-economic and political context (Van Ast and Boot, 2003). Therefore, different public engagement campaigns should not be analyzed only on the basis of the governance style or participant's role, but on their intended outcomes.

In spite of their importance, awareness campaigns provide for limited state–population interactions since participation is normally controlled and limited to target groups. Therefore, public programmes should be encouraged to promote shared responsibility between the government and citizens. This 'citizen-centred' partnership approach helps encourage citizens embrace, adopt and practice a water conservation ethic as part of their lifestyle. It also helps to develop a personal relationship between the individuals and water as one of sustainability's pillars.

At deeper levels of public engagement, governance strategies may focus on participation, delegating, cooperation and facilitation so that participants play a complementary role as advisors, co-decision-makers, cooperating partners and initiators according to their expertise and degree of public involvement. As such, governments generally adopt engagement policies they consider more appropriate according to their governance style. The obvious gain of higher and more meaningful degrees of interaction and participation is a better understanding of the advantages and disadvantages posed by specific situations and, thus, broader public acceptance and support for proposed initiatives. Nevertheless, many governments consider that tighter interaction and participation may not necessarily enhance decision-making quality and, instead, may inhibit rapid policy implementation, which can result in the underachievement of specific goals.

The following is an account of the unique relationship between Singapore's civil society and the government and which is likely to be helpful in understanding the evolution of public engagement campaigns regarding water resources since independence.

Singaporean society

Immigration has shaped the outlook and substance of Singaporean society. Except for Malays, who are an indigenous community, people belonging to other 'races' are either newly arrived migrants or descendents of earlier immigrants. They are mainly classified under four ethnicities, namely Chinese, Malay, Indian and Others, comprising mainly of Eurasians. The population percentages these groups represent remained almost constant during the first 35 years after independence. In 1970, these groups represented 77 per cent, 14.8 per cent, 7 per cent and 1.2 per cent of the population respectively; figures that have changed only slightly, becoming 74.1 per cent, 13.4 per cent, 9.2 per cent and 3.3 per cent respectively in 2010.[1]

Whilst ethnic representation has remained more or less constant, population has significantly exploded and Singapore has grown into a first world city of 5.1 million in 2011 from only 10,683 inhabitants in 1824. At the time of independence, population had already reached more than 1.5 million people, growing at a 4.5 per cent annual rate (Saw, 1991). Back in 1965, average household size for one family nucleus was about six members, gradually dropping to 5.3 in 1977 and 4.8 in 1982. For households with two or more family nuclei (where three or more generations lived together), average size also declined from about nine individuals in 1965 to 8.2 in 1982 (Lee, 1991). Economically, per capita GDP in 1965 was $1,580 (adjusted to 2012 market prices), reaching $7,022 (2012 market prices) in 1977.[2] This inverse relationship between socio-economic development and demographic growth explains why average household size fell to 3.5 in 2001 as income per capita increased to $61,071. That is, average household size declined during Singapore's initial development phase whilst average household income increased almost five-fold in 20 years. [3]

Household income experienced such rapid increments due to the government's efforts to improve growth and development by attracting foreign investment and expertise to pursue rapid export-oriented industrialization as well as by enacting far-reaching labour legislation. One of the main reasons behind the success of these efforts was the strong emphasis the State placed on education. The government saw education as the most important long-term way to improve the quality of life of Singapore's society, to inculcate national values and to train the workforce for maximum economic productivity. In order to reach these goals, and for the first nine years the People's Action Party (PAP) was in power, the government spent nearly one-third of the budget on education to gear academic and vocational preparation more closely to technology and to improve the quality and productivity of the country's workforce (Turnbull, 1989).

Moreover, both economic productivity and education were key mechanisms available to the government to counter challenges posed by political opposition groups and groupings such as unions, clan associations and school systems that had historically shown a tendency to oppose the earlier political colonial elites as well as their successors. The decade following Singapore's stage of self-government in 1959 was dominated by shaky politics, either because of the dispute between two PAP factions over merging with the Federation of Malaya, or because of educational and cultural issues. The newly elected government established an education policy that gave long-term equality in principle to the four main language streams (Turnbull, 2009). It also fostered a multi-racial, multi-cultural and multi-lingual society and promoted respect among all the groups with the overall aim to build a national identity.

In 1957, the country's literacy rate was 52.2 per cent, reaching 84 per cent by 1980 and 96.1 per cent in 2011, gauging the considerable expansion in the country's efforts to promote education.[4] Within 15 years of independence, in 1980, literacy among people aged 15 to 24 reached 97 per cent and 100 per cent in 1992.[5] The number of students was 401,064 in 1962 increasing to 537,278 in 1972. This progress made schools an ideal place to launch public campaigns on multiple aspects such as personal health and conservation of natural resources.

In 1966, then Prime Minister Lee Kuan Yew addressed the severity of divisive cultures during a speech to school principals, when he said:

> Our community lacks in-built reflexes – loyalty, patriotism, history or tradition . . . our society and its education system was never designed to produce a people capable of cohesive action, identifying their collective interests and then acting in furtherance of them . . .
>
> The reflexes of group thinking must be built to ensure the survival of the community, not the survival of the individual, this means a reorientation of emphasis and a reshuffling of values . . . We must have qualities of leadership at the top, and qualities of cohesion on the ground.
>
> (Lee, 1966: 3)

As a result, nation-building and the full development of every child's economically useful capabilities became one of education's main objectives. English remained the state's official language and was increasingly accepted as the development and modernization tongue.

Two short decades after independence, Singapore was physically transformed. Semi-rural areas, dilapidated kampongs, slums and squatter shanties disappeared as their occupants were resettled to the Housing & Development Board's (HDB) newly developed township. New neighbourhoods were equipped with basic amenities and services, including schools, markets, clinics, shopping centres and recreational facilities.

Populations earlier settled along ethnic lines started to live together under the new housing schemes with the objective to achieve racial integration. This made it easier for the government to launch community programmes to improve social interaction and public campaigns on health, environment, family issues, etc. As a result of improvements in the general environment including public housing, provision of water supply and sanitation, education campaigns, etc., health standards rose dramatically. By the 1970s, infant mortality and life expectancy rates compared favourably with most developed countries with 20.5 per 1,000 live births and 65.8 years of age of total respectively.[6] By 2011, people could expect to live up to 82.0 years of age and infant mortality had fallen to only 2.0 per 1,000 live births (this figure refers to resident population not total population).[7]

Like most other institutions, mass media also underwent significant changes after independence. To start with, newspapers in all local languages gradually evolved towards a common outlook with the aim of nation-building. As Chen (1991) mentions, forced by the circumstances, newspapers had to give priority to national objectives, and became instrumental in propagating key national messages and in engaging the population in important matters, such as water. By doing so, they were contributing to the political and industrial stability of the country and thus indirectly to its economic growth.

Since the literacy rate had reached 84 per cent by 1980, there was a greater demand for information and newspaper circulation tripled from 1965 to 1980. By 1976, readership in Singapore had become one of the highest in Asia with 20.9 newspapers per 100 persons (Chen, 1991). Disseminating messages via printed publications thus became an effective way of reaching the population. For instance, Singaporean newspapers have proven to have had a significant role in helping the general public understand intended and crucial messages such as when they gave full support to nation-wide campaigns like 'Keep Singapore Clean' or 'Save water' and more recently presented factual coverage of NEWater. Radio and Television Singapore (RTS) as well as its corporatized avatar – the Singapore Broadcasting Corporation (SBC) – have been instrumental in disseminating educational and awareness programmes among the population on topics such as health, employment, defence and environment to mention only some of them.

Politics and civil engagement

One of the most, if not the most, important and significant political development in Singapore has been the emergence and continuance of a one-party system since the country first achieved self-government in 1959. Political stability, predictability and continuity have ensured economic and social success and allowed the ruling PAP to plan ahead for future challenges by engaging the best and most capable people who would chart the course of the country to greater heights (Lee, 2011).

Chan (1991) mentions that the PAP developed a unique governance style by forming a constitutional representative government, endorsing authoritarian decision-making and concentrating power in a few executives. Such a political decision-making mindset was to be seen in governmental and public activities and initiatives, and consequently, also reflected in the first two public engagement phases regarding water conservation. On their part, Singaporeans were quick to respond to practical programmes but, in general, left initiative to politicians and were content to follow their energetic and dedicated leadership. While most people accepted effective PAP leadership, government's intense activities 'tended to deaden the sense of involvement on the part of the community as a whole' (Turnbull, 2009: 322).

Slowly, by the late 1970s, prosperity and political stability encouraged a more relaxed atmosphere as most citizens were comfortable with their greatly improved living standards. Singapore progressed despite the 1970s oil crisis and the International Stock Market crash in 1987. Consequent elections saw elected opposition members entering Parliament. The government thus adopted a policy of giving some space to the population to express themselves, though non-politically. Feedback Units were established in 1984 and gained momentum with institutions in the form of Nominated Member of Parliament (NMP) and Non-Constituency Member of Parliament (NCMP), for the purpose of hearing views from outside the ruling party in the de facto one-party Parliament (Chong, 2005).

When Goh Chok Tong took over as Prime Minister in November 1990, he looked at wider public consultation and encouraged broader participation of the population (Goh, 1990). As a result, people were given priority as the country's most important resource and the active involvement of citizens was encouraged in different programmes launched by the government. It was also at this time that censorship on art and cinema was relaxed. Slowly, the social environment began to change, and the population at large began to look to voice its opinions but showed little interest in entering the political arena. From 1992, the government began to encourage the public to put forward ideas on environment-related issues and thus the tone and tenor of public engagement campaigns changed accordingly.

Meanwhile, the restlessness of ambitious and well-educated professional Singaporeans began to be expressed in the emergence of civil society groups. The government's idea of these groups was 'a kind of apolitical activism' (George, 2006: 42) that could complement official activities. In this small city-state with limited land and resources and surrounded by the highly sensitive South-east Asian political environment, allowing everyone to sell his/her ideas and putting efforts and resources to realize them, could have its own disadvantages. Therefore, as Ho (2000: 440) mentions, the political system established specific parameters 'that some say' imposes practical constraints on citizens' involvement in policy-making. This struggle has shaped the civil society–state relationship even in present-day Singapore.

In order to learn about the aspirations of the population, the government encouraged 'public consultation' discourses (Chong, 2005: 15). These participatory exercises are also a tool to counter unruly advocacy initiatives, as well as to institutionalize the feedback process, thereby establishing clear frameworks within which advocacy, protest and dissent are de-politicized and staged in a calm and orderly manner. To encourage expression of ideas within law and respected racial and religious sensitivities, an outdoor Speaker's Corner was designated in September 2000.[8]

In consultative mechanisms, the state provides platforms or initiates opportunities where the views and ideas of professionals, experts, relevant social groups and the public may be articulated or heard. This procedure has found its most profound expression in the drafting of Singapore's National Vision Statement, Singapore 21 (Government of Singapore, 1999). Among the several issues Singapore 21 deliberated on was the one on consultation and consensus vs. decisiveness and quick action. In the vision statement, the government encouraged citizens to be active in social services and other civil activities with the purpose of building social capital so that the citizens stayed cohesive and rooted to Singapore.

The committees made for this initiative included Members of the Parliament (MPs), civil society groups, activists, lawyers, unionists, businessmen, professionals, etc., who interacted with the population and listened to their points of view on five main issues: less stressful life vs. retaining the drive; needs of senior citizens vs. aspirations of the young; attracting talent vs. looking after Singaporeans; internationalization/regionalization vs. Singapore as home; and consultation and consensus vs. decisiveness and quick action. Moreover, this public consultation brought some changes in the relationship between civil society and the state as it showed that the government cared about public opinion. For instance, the concept of 'partnership' in public engagement campaigns was established especially in relation to water. And yet, even in initiatives like Singapore 21, the government kept its top-down approach to relating with its citizens.

'Apolitical' civil society organizations that align with the state and its institutions in order to further fulfil the goals and interests of the government are preferred over those who occasionally compete with and challenge the government's viewpoint. For such 'apolitical activist organizations', such as social welfare organizations, the state prefers the term '*civic* society' rather than '*civil* society', 'thus privileging a civic republican notion of citizenship where the emphasis of citizenship was not on individual rights, but civic and national duty' (Chong, 2005: 10). The participation of civic societies in water demand management in Singapore has not been any different and, therefore, their activities have been flourishing under the government's support. A further discussion of their activities is provided in subsequent sections in this chapter.

For an agenda-setting and active decision-making government which prefers civic society organizations and aims at establishing a public partnership to enhance cohesion at grassroots level, but not at the decision-making level, it becomes imperative for its leadership to be strong, visionary and understanding of people's aspirations.

Political leadership

In Singapore, good political leadership has always been cited as the major factor behind its prosperity and affluence. It has been the belief among the elites that leaders are the best judges of the country's destiny and also that the general public's level of knowledge, understanding and interest in political matters do not always enable people to make meaningful decisions about issues that can affect the fate of the country (Ho, 2000). This is a reflection that is also shared by many policy-makers in the world.

So far, this understanding has been vindicated with the impressive transformation of the city-state as well as its continuing economic growth. The fact that the government has been able to instil trust among the population has made civil society bodies and citizens relatively inconsequential in governmental and policy matters. Trust, on the other hand, has been fundamentally necessary to bring effective policy changes through an efficient government, effective legislation implementation, public campaigns, constituency work, efficient performance of its utilities and other government agencies, and public campaigns.

It is again to the credit of Singapore's leadership that despite its elitist policy-making approach, it has always remained open to alternative course corrections. Various mechanisms like Feedback Units, Service Improvement Units, Citizens' Consultative Committees (CCC) and various meeting platforms with elected representatives, have opened up opportunities to citizens seeking to influence policy inputs. In the early 1960s, PAP leadership activated grassroots institutions to improve the linkage between the government and the population. Community Centres and People's Associations (PA) were made the focal point for community integration through vocational, recreational and sports activities. They were seen as a catalyst for generating a cohesive sense of community at the constituency level (Lee, 1978). CCCs were also developed to transmit government policies to the population and to relay grassroots demands back to the government. CCCs enhanced the capacity of political leaders to govern, also undertaking the major responsibility of managing and conducting multiple official and educational campaigns at the grassroots level designed to re-orientate the behaviour of citizens (Chan, 1991). In the later years of the 1970s, Residents' Committees were formed in all HDB estates to foster community cohesion with the government ensuring the involvement of middle and senior level administrators within the Committees. Highlighting the importance of this involvement, former Prime Minister, Goh Chok Tong, explained:

> Senior civil servants must get to know the aspirations, needs and feelings of the people in order to perform their duties more effectively. By making them sit on RCs to expose them to grassroots problems, they are given the opportunity to demonstrate their ability at problem solving and establishing rapport with residents from all walks of life.
>
> (*The Straits Times*, 4 April 1980)

In Singapore, the shortage of public forums and civil society organizations has made the role of MPs even more relevant, who have traditionally acted as important mediators between the government and the population. In addition, parliamentary debates have also been a source of public policy education as these have been publicized for decades on television, radio and in newspapers.

One could say that, compared to their task in Parliament, the role MPs play in their constituencies is more important from the public awareness point of view. The weekly 'Meet the People' sessions MPs are expected to attend are fundamental for close government–people contact as they provide the opportunity to develop crucial grassroots links. Through these meetings, leaders learn to accurately assess the impact of government policies and the way the population reacts to them. Generally, MPs remain in close contact with local community leaders who have been co-opted into the CCCs. While MPs try to attend as many CCC meetings as possible, citizens' committee members help MPs explain policies to the general population.

The Public Utilities Board (PUB) has dedicated staff for monthly MPs' meetings at the local level, irrespective of whether water is an agenda item. The objective has been to resolve any water-related problem that may be raised. MPs are also invited to attend the PUB programmes organized in their constituencies.

MPs are also expected to assist in the implementation of national goals in their respective constituencies and organize the corresponding campaigns. For example, from its very onset, the Singapore government has launched numerous campaigns to emulate particular practices and behaviours it has considered attractive, useful and relevant. The 'Anti-Spitting Campaign', 'Keep Singapore Clean and Pollution Free' and the 'Save Water Campaign' are the earliest examples of activities organized by local MPs, in cooperation with the CCCs and the constituency's branch (Chan, 1976). As in the past, MPs continue being instrumental in mobilizing people to participate in water conservation campaigns.

In the following various strategies, campaigns and programmes launched by the government along the years to engage the population in water conservation are discussed in detail. Overall, during the post-independence days, Singapore was mainly engaged in nation-building and public agencies looked for professionalism and behavioural change among their employees. Information and awareness campaigns encouraging nationwide cleanliness began in the early 1970s making use of posters, pamphlets and other didactic material as well as programmes broadcasted on radio and television.

Launched in 1977, the 'Cleaning of Water Courses' project proved highly motivating despite being top-driven and it made people immensely aware of the need to keep the water bodies clean. Continuous national water conservation education and information campaigns did succeed in creating public awareness to save water. Equally, concerted efforts to educate students on water issues further helped inculcate water conservation habits.

In 1992, consultative awareness programmes were launched with mass-scale campaigns and broad consultations with people from all sectors. By 2001, when the PUB was reconstituted as the national water agency and it adopted a new set of corporate values, it began to design diverse, interesting and innovative engagement campaigns to attract popular interest. It thus adopted a consultative participation framework that used interactive activities to actively engage the population. The breakthrough came in 2004 when the general public was encouraged to carry out activities near to or at water bodies so that people would develop a personal relationship with water. Such a scheme was looking not only for public engagement but also to build long-lasting partnerships. Examples include the 3P strategy (a People, Public and Private sectors partnership) and campaigns such as 'Water for All: Conserve, Value and Enjoy'.[9]

Public relations activities

The story of public relations in water services started on 1 June 1964 when a Public Relations unit was set up as part of the PUB's General Manager Department. It had the responsibility to promote good relations between the consumers and the Board, attending public complaints and enquiries, advising on the services provided and liaising with the heads of departments concerned (PUB, 1964).

In the beginning, interaction with the public was limited to responding to complaints, offering employment news and disseminating information related to the PUB's activities. The PUB Newsletter, which started in December 1964 with a circulation of about 2,100 copies, disseminated information on the utility's identity and goals; employment opportunities; the technological and staff profile of the PUB employees; progress achieved in solving problems as well as in addressing complaints, etc. (PUB, 1965).

In March 1966, the PUB-related information was featured in four languages over radio and television for broader dissemination. In an attempt to further improve the Board–consumer relationship and enhance its public image, the PUB conducted 'Courtesy Talk' sessions on public relations for a total of 190 officers who were in contact with the public, including cashiers, counter clerks and meter readers (PUB, 1966). This trend continued for several years with the public receiving information on the PUB's roles, responsibilities and accomplishments as well as major policy decisions and approved schemes through press releases (PUB, 1967). In 1969, the PUB started disseminating messages to discourage water wastage and, that same year, the agency actively participated in the government-organized 'Keep Singapore Clean' campaign. The agency then borrowed this concept and applied it to 'Keep PUB Clean' seeking to maintain and encourage higher standards of cleanliness in the Board's buildings and quarters throughout the city-state. The PUB organized competitions among its employees for the cleanest quarters, offices and grounds with more than 100 senior officers and Board members volunteering to clean the stretch of beach facing the Board's holiday bungalows at Tanah Merah Besar. Improvement campaigns were also launched in and around the PUB premises and periodic inspections were conducted to ensure they were kept clean (PUB, 1969). So far, however, population was not included in the PUB's cleanliness campaigns since these addressed only staff members and their families (PUB, 1970). These internal efforts were mirrored by public-oriented initiatives.

By this time, academia also started raising its voice to denounce water wastage and stressed the need for its conservation. These facts were well recognized by the PUB, who also used to issue warnings to the population to conserve water in order to avoid rationing. Concerned by the situation and aware of the importance of making people aware of water shortages, newspapers played a vital role by including headlines such as 'when will the cuts begin?' almost on a daily basis, describing it as a 'national problem' and urging people to save water (Chia, 1971a and 1971b; *The Straits Times*, 9, 10, 12 and 18 May 1971). *The Straits Times* also initiated the 'Don't Waste Water' campaign and published daily stories on water consumption for over seven years from 1971 to 1978.

Such initiatives were motivated by the 1971 prolonged dry spells, which had also made the PUB issue media statements on water conservation:

> The spectre of water rationing is going to be with us for the next few months . . . If the public does not continue to maintain their efforts in conserving water and reduce further their daily consumption, there is every likelihood that we will not be able to tide over the period without rationing.
>
> (*The Straits Times*, 16 May 1971: 1)

That year's dry period lasted longer than expected, mobilizing other groups to join the public appeal to use water wisely. For example, the Singapore Private Market Owner's Association urged its members to save water and pinpointed certain wasteful practices that ought to be discarded:

> From experience, it has been the practice of poultry stallholders to wash their poultry and pluck the feathers from a tub with water running all the time. They are the worst offenders, and vegetable stallholders rate second, as they wash their vegetables under a running tap. Next comes the fish mongers. These three categories are the worst for wasting water. If they can realize the value of water and the inconvenience to them and others, if rationing is imposed, at least 500 gallons from each market per day can be saved, thus helping to bring the target figure nearer and may avoid water rationing.
>
> (*The Straits Times*, 28 May 1971: 8)

The seriousness of the 1971 drought can be judged from the fact that during the first four months of the year, rainfall was less than what it had been for that same period in 1963, when rationing had been enforced (*The Straits Times*, 9 May 1971). Wide-reaching efforts to incentivize better water usage paid off. The PUB campaigns had a good public response as daily water consumption was reduced substantially. According to newspaper reports, within 10 days (15 May to 24 May), daily water consumption was brought down by 20 per cent (*The Straits Times*, 18 May 1971: 1; 25 May 1971: 1).

Nonetheless, since daily consumption was still not within the target level, the PUB issued a warning stating that those who wasted water would be fined $500 (*The Straits Times*, 28 May 1971). Thereafter, the PUB inspectors started imposing penalties on individuals who had been found wasting water and the media helped denounce, deter and expose such behaviour by publishing a myriad of related stories.

Combined efforts proved successful since daily consumption in June 1971 remained mostly below 100 gallons compared to the peak consumption of 130 gallons/day in early May 1971. Moreover, as scarcity continued, the PUB's instrumental experience in dealing with droughts was once put into use during the 1972 water conservation campaigns. That same year, the Board also replaced the use of imperial gallons to charge for water consumption and introduced the metric system (up until that time, fees were based on 1,000 gallons).

Beginning of public engagement campaigns

As mentioned earlier, water scarcity and dependence on imported water was slowly becoming a reason for concern as average daily consumption had increased from 81.9 Mgal in 1965 to 113.8 Mgal in 1972 (PUB, 1965, 1972). Irrespective of continued reminders to the population and industries on the importance of water conservation and of relative success in improving water demand management, 1972 was a year of high water consumption. Domestic consumption was 10.38 per cent higher than the previous year and industries were using 20.93 per cent more water compared to 1970. Total annual water sales rose from 147.8 Mm^3 to 159.4 Mm^3 (Ooi, n.d.). Such spikes became a reason for political preoccupation, especially since the country remained dependent on external sources for its water supply, compelling it to begin looking for alternative water sources and strengthening water conservation efforts. Tan *et al.* (2009) mention how both the HDB and the Ministry of the Environment (ENV) embarked on a pilot scheme to supply industrial water for toilet flushing to housing estates in order to replace the use of potable water for non-potable ends. More importantly, high-quality, purified and reclaimed water was given serious consideration as a potential alternative to increase the amount of water supplied to the system. However, at this stage, it was urgent to raise public awareness on water conservation.

A national water conservation campaign was launched by the PUB in November 1972. Under the 'Water is Precious' slogan, the campaign sought to make people aware of the growing importance of water and to inculcate and encourage water saving habits. Community activities were organized around the island to show people the many small practical ways to save water and highlighted the practical dos and don'ts (PUB, 1973). The PUB used press, radio and television to publicize the campaign among the population, which became the Board's first large-scale consumer-oriented campaign. Underlining the emphasis on water saving, the then Minister for Education and PUB Chairman, Lim Kim San, stressed that 'to save water only when there is a drought is not good enough for us in Singapore. Saving water must become a daily habit with us' (PUB, 1965: 3).

By 1973 water conservation actions began to bear fruit, and for the first time since 1967, Singapore witnessed a negative growth in domestic water consumption (Ooi, n.d.). School children were identified as a major target group and ever since these early initiatives, efforts to educate them on water conservation and to inculcate good water use habits have been maintained and continued. In addition to public campaigns in the mass media, consumption was also brought down through tariff mechanisms that increased prices for domestic users.

Once again, after a sudden drought in 1976, the PUB published appeals to large water users, such as hotels, coffee shops, laundries, etc., to avoid wasting water (PUB, 1976). Slogans such as 'Don't wait till the last DROP – Save water now' were displayed in public places, publicizing the need to save water. The 'Water is Precious' exhibition was revived that same year. Ministries, agencies and associations including the Prime Minister's Office, the Ministry of Culture and the PA collaborated and together toured 12 community centres to make a success of the exhibition. The campaign's message was disseminated through posters, leaflets, stickers, outdoor projections and sponsored programmes.

At that time, water bodies were kept strictly out of bounds for the population. One example is the response from the PUB to the public requests to allow fishing in the reservoirs in the same way that it was allowed in other developed countries:

> conditions are different in these countries primarily because their reservoirs and catchment areas are relatively larger in relation to the population density. The pollution load caused by anglers in considered negligible in such reservoirs . . . in our case the most serious problem of pollution is posed by the activities and presence of a large number of anglers.
>
> Even with the present restricted activities in the reservoir grounds litter is frequently found in the reservoir water. Littering apart, there are health considerations which make public fishing in our reservoirs an undesirable activity.
>
> (Vaz, 1977: 5).

Singapore river cleaning operations

The year 1977 began with the inauguration of the impounded Upper Peirce Reservoir and set the pace for the myriad of redevelopment activities that would follow in the coming years. Prime Minister Lee challenged Singaporeans 'to keep every stream, every culvert, every rivulet free from unnecessary pollution' and adopt a way of life that would keep waters clean (Ministry of the Environment, 1987: 8). In consultation with the PUB, Television and Radio Singapore produced the documentary film *The Search for Water*, highlighting the Board's efforts in its continual search for new water resources (PUB, 1977).

In 1978, the PUB adopted a new slogan: 'Adapt, Innovate and Prosper', perhaps a timely indicator of change both on approach and attitude. Lee Ek Tieng, who was heading the Singapore River cleaning operation at the time, was appointed PUB Chairman in 1978, with which the year saw the beginning of publicity programmes to raise greater public awareness about the agency's roles and responsibilities. This initiative put in place a series of regular features that remain up until this day, such as arranging visits for new employees, students, organized groups and overseas visitors to tour water treatment works and power stations. The first activity of these public outreach programmes was the official opening of the Electricity Department's Jurong Depot Complex for public visits (PUB, 1978).

As yet another initiative to develop better public relations between consumers and the Board, MPs were invited to attend a briefing session on the Board's policies and operations. Later on, three more briefings were held for representatives from community centres and CCCs and officers from the PA. These activities proved very useful to gather suggestions, solve problems and clarify doubts (PUB, 1978).

Liaising with consumers

As has already been noted, 1978 saw the implementation of a myriad of initiatives to raise water's profile among the population. That year, some 8,163 individuals participated in the Meter Reading Contest as well as in the many other activities organized for consumers to grow awareness of the importance of water and to teach them to read the meters on their own. A bilingual audio-visual programme called 'A Guide to Meter Reading' was screened at community centres, schools and other public places. One year later, and seeking to bring people closer to water and provide more recreational facilities, an initiative to open up Kranji and Upper Pierce reservoirs was launched (PUB, 1979).

As conservation efforts intensified, rising population and the rapidly growing industrial sector continued to sharply increase water consumption. In less than a decade, overall water consumption in Singapore rose by 46 per cent between 1972 and 1981. The domestic sector alone saw an increase in consumption of 51.2 per cent in 1981 from the 1972 level (Ooi, n.d.). With such spikes, it thus became imperative to deliver the conservation message to consumers through more effective means other than information. In 1981, the PUB undertook the task of carrying out several concurrent initiatives. To start with, the Water Conservation Unit was established to promote water conservation in the domestic and non-domestic sectors as well as to suggest suitable policies. One of the unit's roles was to liaise with large industrial water consumers and suggest to them how they could reduce water consumption. This information exchange, which became a regular feature in the coming years, also facilitated taking feedback back to the PUB. That same year in September, and as part of the implementation of a Water Conservation Plan, officers from the Water Conservation Unit visited some 4,000 domestic and non-domestic consumers. A further 640 commercial and industrial consumers responded positively to the PUB's advice to appoint water controllers to monitor water consumption (PUB, 1971).

. . . and with population

To draw public attention to the importance of reducing water wastage and water consumption, a one-month long 'Let's Not Waste Precious Water' campaign was launched in addition to the talks the PUB officials conducted in schools, colleges and vocational institutions to encourage water saving measures and reducing wastage (PUB, 1981). The result of all of these persuasive water conservation measures, engaging the public through information, awareness and education, was that average water consumption grew slightly less than the previous year but remained significantly high. In spite of these efforts, revising water rates was considered as the best alternative to induce more efficient water use and to reduce wastage. *The Straits Times* published the PUB's proposed tariff revision strategy and welcomed it as a measure seeking to curtail the 7.7 per cent water consumption growth rate that had been recorded in 1980 (*The Straits Times*, 30 July 1981).

Newspapers have been extremely supportive of the PUB's save water campaigns. In one of its editorials explaining Singapore's water needs and role of the PUB, *The Business Times* wrote:

> Seen in this light, the actions and directions adopted by the Board become more understandable. Unless consumers are able to economise on water consumption on their own, they should welcome the PUB's efforts to help them curb wastage. After all the Board has gone out of its way to spend time, energy and money by sending its officers to households and establishments that appear to be using water excessively to help them find means by which they can save water and money.
>
> However reflecting the rather complacent attitude of many Singaporeans, the response to the PUB's overtures of assistance was cool. But, undaunted by this, the Board has decided to try more earnestly to make habitual wasters mend their ways. The trouble that the PUB has taken to provide free advisory services surely deserves a more positive response from the big consumers.
>
> (4 November 1981: 8)

Institutional anniversaries also became an opportunity to be seized to further disseminate the water conservation message. In 1983, the PUB's twentieth anniversary was harnessed to organize various consumer visits to the PUB installations to provide information about its activities and create awareness about the importance of saving water. The response was enthusiastic with some 37,000 people visiting various PUB facilities that year (PUB, 1981).

Further initiatives

Similarly to the 1981 one-month long 'Let's Not Waste Precious Water' campaign, national water conservation campaigns were organized from 1983. These were launched to raise greater awareness among the public and industry and mainly focused on commercial and industrial sectors since industrial water consumption

had increased from 29 per cent of total consumption to 36.4 per cent within a decade (1973 to 1983). By the third quarter of 1983, the Water Conservation Unit
had inspected 7,500 large water consumers' premises and persuaded more than
70 per cent of them to adopt water conservation measures. As reported by the
PUB, these efforts reduced their annual water consumption by 11 per cent (PUB,
1983). To accompany and reinforce the achievements of the water-saving campaigns in the non-domestic sector, the government began providing incentives
to those consumers who succeeded in cutting down their water use. For example,
a 50 per cent investment tax allowance for water conservation equipment was
announced for those industries that substantially reduced their water use (PUB,
1983). In 1984, as a result of the awareness visits to promote water conservation
held for private organizations and enterprises, more than 700 commercial establishments installed water-saving devices such as self-closing delayed action taps
and constant flow regulators (PUB, 1985a). Industries were also encouraged to
use industrial water instead of potable water.

In 1985, in addition to the previously used engagement methods such as exhibitions, seminars, logo designs, competitions, talks and frequent visits by Board's
officials, community leaders were also encouraged to get acquainted with PUB's
activities and on the importance of water conservation (PUB, 1985a). One year
later, in 1986, a different method was used to disseminate water conservation
messages and educate the public on the steps to be taken in case of an emergency.
The Singapore Civil Defence Force, several grassroots organizations and approximately 3,500 households participated in an 'emergency' water exercise (PUB,
1986). This activity was meant to target the population born after independence
and who had never experienced a water crisis in their life. The net result was that
domestic consumption in 1986 dropped by 2 per cent from a 4.3 per cent increase
in 1985 (Ooi, n.d.).

In any case, it was considered that water conservation campaigns required a
further boost through education, keeping in mind that this could have a long lasting impact on young minds. In 1987, a water conservation course was introduced
at secondary level, helping 596 students understand Singapore's water challenges
through 18 organized sessions (PUB, 1987). In 1988, a further 3,800 students
took part in the course and many more in subsequent years (PUB, 1988). Educational kits were also published and distributed at schools as these remained a focus
of public education campaigns in the 1990s and beyond.

Over those same years, an unusually long dry period forced Singapore to once
again make public appeals to cut down on unnecessary water use by 10 per cent
and conserve falling reservoir stocks (*The Straits Times*, 7 April 1990). The then
Minister of State (Foreign Affairs and Finance) George Yeo publicly emphasized
the following message: 'Water is precious and a matter of life and death . . . Singaporeans should inculcate in themselves a consciousness that water does not come
by easily' (*The Straits Times*, 9 April 1990).

Extensive and intensive campaigns bore fruit and, in one short month,
from March to April 1990, daily water consumption in Singapore dropped by
11.2 per cent. This improved the level of reservoir stocks and it was no longer

necessary to ration water that year (PUB, 1990). Once again, newspapers played their part during this period. They published stories revisiting the 1960s drought and rationing, reminding people of the difficulties faced during that period and urging them not to waste water (*The Straits Times*, 22 March 1990; Tan, 1990).

Nevertheless, once the dry season was over, and despite the extensive use of information, awareness and education campaigns as tools for water conservation, domestic as well as non-domestic consumption increased once again. Therefore, to reinforce the various voluntary engagement approaches, in 1991, a Water Conservation Tax was introduced as a pricing tool to discourage excessive water consumption. Under this mandatory requirement, a 5 per cent tax was to be levied on monthly domestic water use above 20 m^3, and a higher 10 per cent rate would apply to non-domestic consumption. In spite of this economic measure, and its subsequent increases in 1992 and 1995, water consumption in those years increased annually more than 3.99 per cent for domestic use and more than 6.2 per cent for commercial use (Ooi, n.d.).

Outreach and education programmes

At the rate water consumption was increasing, there was the risk that it would double every 16 years. With this in mind, and taking into consideration the efficacy of the different public engagement methods used until then, the PUB now focused on even wider reaching mass-scale conservation campaigns that engaged entire families. This approach was encouraged by the Singapore Green Plan launched in 1992, which charted the strategic directions that Singapore would be adopting to achieve its goal of sustainable development.[10]

The Singapore Green Plan demonstrated that effective public participation was key in achieving future policy targets. Tan *et al.* (2009) mention how the ENV realized that for the Green Plan to be effective, it had to resonate with the people for whom it was intended. Therefore, public fora were held to explain the plan's goals and targets and further publicized it, gaining important popular support and feedback with more than 100,000 people visiting the Singapore Green Plan exhibition in 1993. The extensive public consultations the plan encouraged inspired policy-makers to emulate a similar type of engagement process for other campaigns as well. This was the beginning of 'public consultations' as an engagement tool in environmental matters in Singapore.

With the active support of teachers, large-scale engagement campaigns were extended to a number of schools. More than 20,000 students participated in the PUB's talks on water and energy conservation in 1993 alone (PUB, 1993). Similarly, in 1994, about 12,000 students were educated on the importance of using water and energy efficiently (PUB, 1994). The Board's Energy Conservation and Exhibition Centre continued offering regular courses on water and energy conservation, reaching out to 10,798 students in 1995 (PUB, 1995). Moreover, since 1994, the National Environment Agency (NEA) started a Network of Environmental Education Advisors (EEAs) as a communication platform between teachers and the agency. This programme was aimed at changing students' behaviour from a young age and inculcate in them sustainable resource conservation. Another effort includes community involvement programmes organized by the Ministry of Education, which encouraged students to join community activities related to water, resource conservation, public hygiene and recycling, for instance (Tan *et al.*, 2009).

Messages were thus disseminated and reinforced using both conventional information tools and more audience-friendly approaches using interactive radio programmes. In 1995, the PUB launched yet another 'National Save Water Campaign' to raise awareness on water scarcity and the need for the conservation of the resource. The initiative mainly targeted community groups and students, and for the first time, major shopping centres staged campaign exhibitions to more widely disseminate information about the need to conserve water among the general public. The same year, in an unusual campaign that ran for six days, an island-wide water rationing exercise was conducted involving 30,000 households. During this period, water supply was interrupted for 14 hours on each exercise day (PUB, 1995). The aim was to shake up public inertia, especially among youths, and remind Singaporeans about the importance of water.

All previously mentioned initiatives targeted mainly Singaporean nationals but as the country industrialized, the number of foreign workers had increased substantially since the 1970s. By 1995, there were about 350,000 labourers from abroad, constituting 20 per cent of the sector's total workforce.[11] That year, and for the first time ever, foreign workers were targeted by water campaigns, and leaflets in Thai, Singhalese, Tagalog and Bahasa Indonesia distributed among them explaining how to use water wisely (PUB, 1996).

In 1997, the PUB implemented sustained and continued outreach and education programmes to reinforce the visibility of water campaigns. For example, a new Water Conservation Centre was established in Bedok offering several interactive exhibitions (PUB, 1997a). A new water tariff regime was set the same year to introduce greater fiscal incentives for customers who used water rationally, and following the series of recommendations and suggestions proposed in the many information campaigns. It was thought that higher water prices would draw attention to water management's need to 'conserve' and 'value' the resource. These fiscal measures were given timely support through massive conservation campaigns.

Several enthusiastic groups were also established for environment protection purposes. The process was slow and mostly inspired by the Singapore River cleaning but it bore fruit. In 1997, a modest advance was made by the Waterways Watch Society (WWS) to meet the challenges of maintaining the scenic river conditions at Boat and Clarke Quays that emerged from cleaning up the Singapore River. Within a year, the Society collected 4,000 items of litter during its water patrols. Since then, about 60 of its members have been constantly picking up litter along the waterways in their boats. As part of the long-term strategy to keep waterways clean through education, the Waterways Watch Society has also been working on water sports activities and events like the 'Clean and Green Week' and the 'International Coastal Cleanup' (PUB, 2004).

In 1998, awareness programmes were taken a step further. The ENV, along-side grassroots organizations, launched a community-engagement programme providing residents with a platform for information dissemination on dengue prevention. The first Dengue Prevention Volunteer Group (DPVG) was formed in Serangoon Gardens in 1998. By 2009, the number of volunteers joining these types of initiatives had increased to 6,200 (Tan *et al.*, 2009). This effort triggered voluntary communal interest and commitment towards different water aspects and public health issues.

From 2000 on, numerous water conservation commercials were screened on electronic billboards and 'Save Water' messages broadcast with the support of Singapore's Telecom. Similar messages were also printed on the transportation system Transit-Link fare cards. These ubiquitous messages were planned to be constant reminders to the population on the importance of water conservation. Furthermore, more direct measures were also taken to manage water demand. The successful experimental use of thimbles to reduce water wastage inspired the Singapore government to send such devices to every household.

PUB's reconstitution

On 1 April 2001, the PUB became Singapore's national water agency and was transferred from the Ministry of Trade and Industry (MTI) to the ENV. The Board's new avatar came with a new set of corporate values and with a mission to secure an adequate supply of water at an affordable cost. Five broad strategies were identified to fulfil this mission, including yield maximization and diversification of water resources; water reclamation and reuse; wastewater proper treatment and disposal; and storm water management and water demand management (PUB, 2001). Ever since, public engagement has remained an important tool for demand management and the PUB has continued with programmes such as conservation talks in schools, incorporating water conservation messages and tips in school textbooks and children's publications, distributing water conservation material among the general public and staging water conservation exhibitions.

Efforts to achieve public acceptance of NEWater

In contrast to various disputed US and Australian projects involving reused wastewater, in Singapore, public acceptance of NEWater has been a smoother process. More than three decades of sustained and continuous water conservation campaigns have helped the local population understand that recycled water has to be one of their sources.

Being a small city-state has helped Singapore in its successful implementation of policies, and also in making the population aware of the importance of water. As such, the island's reduced size has saved it from problems arising from multi-tier governments, complex regulatory arrangements and overlapping authorities and responsibilities. The PUB has shown dynamism to outsmart the prevalent conservatism in most of the water utilities around the world. Singapore's dependence on imported water and a limited land area has made the Board particularly proactive in improving its management practices, establishing suitable pricing mechanisms and also in embracing the most modern technology available.

That cutting-edge technology has been put behind NEWater. It was only after more than 20,000 comprehensive tests and analyzes that NEWater was certified as a safe and sustainable water supply source and production began in 2002. In September that year, the government of Singapore decided to add 2 Mgal/day of NEWater (less than 1 per cent of the island's daily consumption) to its reservoirs. In 2012, NEWater has been able to meet 30 per cent of total water demand and this share is expected to reach 50 per cent by 2060.[12]

After NEWater was successfully launched in 2003, the PUB adopted an information campaign focused on the Four National Taps as the city-state's water supply strategy. Singapore's own water catchments to harness water supply constitute the first tap; water imports from Malaysia represent the second tap; NEWater stands as the third tap; and desalinated water has become the fourth national tap. To make Singapore's water supply system dependable and resilient, this source diversification, along with efficient demand management, was considered as a crucial element. This was not only a confident declaration of Singapore's capabilities on water supply to its population, but also a subtle attempt to make people accept NEWater as a permanent water source.

The NEWater Visitor Centre, which opened in February 2003, has become the focal point of the PUB's public education on NEWater. The centre highlights the importance of water and how Singapore leverages technological advances to reclaim the resource. Visitors are able to view first-hand the operation of the advanced dual-membrane and ultraviolet technologies used to produce NEWater. As of 2011, it had attracted close to 800,000 domestic and foreign visitors (PUB, 2011a).

To overcome the psychological apprehension of people in accepting NEWater, deliberate attempts were made to shift public attention away from the source by focusing instead on the treatment process involving state of the art membrane technology. The PUB consciously avoided using terms carrying negative connotations and replaced them with positive ones. For example *wastewater* and *sewage* were dropped and the new term *used water* was coined (Leong, 2010).

The Board also began an intensive publicity programme on the importance of NEWater, its quality and production processes. Mobile exhibitions, briefings, posters, advertisements, brochures and public talks were used extensively to educate the population about NEWater as a safe and sustainable drinking water source. Extensive public campaigns resulted in widespread support and within six months of being launched, 1.5 million bottles were produced for distribution in community events. During the 9 August National Day Parade in 2002, the Prime Minister, other senior leaders and 60,000 people attending the event toasted with NEWater (PUB, 2002a). To show that people had embraced NEWater, PUB produced and distributed five million bottles to the community for public sampling in the three years after it was first produced (PUB, 2005b). The same year, the PUB launched PUB-One, a contact centre that has since then acted as a single window available to customers to put forward their enquiries and provide feedback on issues related to water, sewerage, drainage and NEWater round the clock. To do this efficiently, six multi-channel contacts have been established (PUB, 2002a).

Media management

Media in Singapore has been very supportive in reporting issues related recycled water. As Leong (2010) mentions, between 1997 and 2008, more than 200 media notes on recycled water were published in the region's three major newspapers (*The Straits Times, The News Paper* and *The Business Times*). Most of the reports carried positive tones or favourable opinions about recycled water and only a few gave a negative impression or conveyed unsatisfactory opinions about recycled water, which had been mainly expressed by neighbourhood politicians.

The attitude taken by the Singapore media was rational and very construc-tive. Media believed that it was important for readers to understand how critical it was for Singapore to have clean water supplies and also to be informed on the safe quality of NEWater. In contrast, media's reaction has been very different in places such as Queensland, Australia. From 2005 to 2008, when the province became serious about producing recycled water after a prolonged drought, almost one-third of all media reports published at that time carried negative comments on recycled water (Leong, 2010). Even the language the media used to refer to recycled water differed in the two cases. Whereas in Singapore, the media stressed how recycled water was *cheap* (less costly), *purified* and *tried and tested*, the Aus-tralian media used terms such as *treated effluent, toilet to tap* and *shit water*.

The PUB has been responsible for efficient media management. Not only regu-lar media briefings were conducted at the time NEWater was being launched, but reporters were also invited to visit places where water reuse projects were running successfully. This was instrumental in making the media realize that used water had been utilized for many years in some areas of the world. Equally important was to convince religious leaders about the reliability of the advanced technol-ogy. The PUB encouraged them to test NEWater's quality for themselves and explained the importance of considering the water cycle rationally and logically rather than emotionally. It convinced them that, in fact, there is no such thing as 'fresh' water and that nature has recycled water naturally countless times.

The 3P partnership and the new age of public engagement

With more than half of Singapore's land also acting as a catchment area, it became important for the PUB to make the population aware of the fact that they were living in water catchment areas. Also, that the rain falling in their neighbour-hoods was collected, transported to the reservoirs and then to the treatment plants before distributing it to their homes. The PUB thus decided to make attempts to increase people's understanding of this 'cycle'. The objective was that once the population understood the message, they would be more water responsible.

The idea of promoting activities near the reservoirs was contrary to the limitations imposed in the past. This was a paradigm shift from the times when littering, industrial pollution and silt discharge represented real problems that prompted the authorities to ensure people stayed away from reservoirs. With the adoption of advanced treatment technology and the premise that people carrying out activities in the water would be the last to pollute it, the PUB began planning the development of a personal relationship between the general population and water bodies (PUB, 2004). A communication strategy was thus elaborated to disseminate messages in more subtle and emotional ways but with permanent impact. Water was to be made attractive to draw public interest and a proactive media approach would be adopted to raise awareness about the PUB's activities.

In 2004, multiple activities encouraging people to enjoy water and develop a relationship with it were introduced in what came to be known as the 3P (People, Public and Private) approach. The rationale behind this network is to manage water demand by building people's affinity with water so that they keep it clean and gradually take stewardship in its conservation. A noticeable term used in this approach is 'partnership', as it looks to gain people's support and that of the public and private sectors as well with the objective to fulfil water demand management goals. Partnerships indicate pragmatism and contributions from the different stakeholders based on their capacity, capability and commitment.

Water conservation in the domestic sector

Despite NEWater's successful launch and the commission and opening of NEWater factories at Kranji and Seletar, the PUB has continued focusing on water conservation. A Water Efficient Homes (WEH) programme was put in place in 2003 to help households save water and under which advisors and grassroots organizations were expected to lead their respective constituencies to have water efficient homes by installing water-saving devices and adopting water-saving habits. By the end of the year, 100,000 homes had been fitted with water-saving devices that were contributing to 5 per cent water savings in their homes. According to the PUB, this contributed to the slow 0.5 per cent increase in domestic water consumption registered in 2003 and the maintenance of the 165 l/capita/day domestic water consumption for the fifth consecutive year (PUB, 2003).

Further, to reinforce its efforts to bring per capita consumption down to the 160 l/capita/day mark, and as part of the WEH programme, the PUB and grassroots volunteers distributed more than 333,000 water-saving kits. Under this scheme and to reduce excessive water flow rates, households were encouraged to install water-saving devices such as thimbles on taps and showerheads and water-saving bags in 9-litre flushing cisterns. The PUB also provided the necessary technical and logistic support for the programme. The 'Do-it-Yourself' kits comprised water-saving devices, installation instructions and water conservation tips were supplied free of charge. A brief demonstration exhibition was set up in each constituency for residents to address any doubt on how to operate the devices and adopt the programme's suggestions (PUB, 2003).

In 2006, the '10-Litre Challenge' was launched to encourage every person to reduce daily water consumption by that amount. Initiatives include a web portal to educate Singaporeans on the different ways they can save water at home and better follow the WEH programme. What is more, the PUB officers and volunteers have formed Water Volunteer Groups (WVGs) to conduct visits to households, including private housing units, recording higher water consumption to help them with the installation of water-saving devices and share water-saving tips.[13]

As part of its ongoing efforts to raise awareness on the importance of using water wisely, the PUB introduced more targeted activities to reach out to different segments of the population in 2011. These initiatives were aimed at encouraging the public to make water conservation practices a way of life and included television advertisements, water audit projects by students, water conservation training by employment agencies recruiting domestic help, revamped water-saving kits and roving exhibitions. Following the successful implementation of the awareness programme in 2011, the PUB's efforts continued in 2012, this time disseminating the message through social media.

Water conservation in the non-domestic sector

Besides good management and maintenance practices, flow rate control is also a key factor in managing water consumption in the non-domestic sector. A PUB-conducted pilot project study revealed that a 2 l/minute flow rate in public taps was sufficient for normal washing purposes. Based on this finding, non-domestic customers were advised to review and reduce water fitting flow rates in their public and staff toilets. The PUB's 2003 Annual Report touched on the encouraging response from public agencies, hawker centres and town councils that resulted in up to 5 per cent savings in monthly water consumption that year. By March 2004, the PUB had also launched the Water Efficient Building (WEB) programme to encourage building managers and owners to make their properties water efficient. Since it was launched, more than 2,200 commercial buildings adopting good water measures have been WEB certified.

Throughout the years, water-saving initiatives have grown in inventiveness and resourcefulness. In 2008, the PUB started an umbrella programme for getting the non-domestic sector to improve water efficiency and reduce water consumption called the '10% Challenge' programme. It focuses on raising awareness and building businesses' capacities to improve water efficiency. Its portal helps to provide a complete assessment of the current status of an organization's water demand, benchmark itself against similar organizations, conduct audits to identify opportunities for improvement and lists a series of water efficiency measures for different building types.[14]

Moreover, the PUB has taken to the task of working with industries, hotels, schools and hospitals in the preparation and voluntary submission of Water Efficiency Management Plans (WEMPs). The scheme analyzes current water use, it identifies potential water-saving actions and provides an implementation timeline for such measures. The PUB's existing Water Efficiency Fund can also be used to finance WEMP-identified measures, if customers meet established criteria.

Over the years, the PUB has developed quite a large number of innovative initiatives to promote education, raise awareness and disseminate information on water conservation and efficiency practices for both the domestic and non-domestic sectors. Examples include, but are not limited to, the water efficiency labelling scheme (WELS) (jointly developed with the Singapore Environment Council), the mandatory water efficiency labelling scheme (MWELS), the Water Efficiency Fund, the Water Adopters programme, WaterHub (aiming at bringing the water industry together and provide opportunities for technology development, learning and networking) and the formation of the Singapore Water Association to foster dynamic collaboration among private sector players.

The 'Active, Beautiful, Clean Waters' (ABC Waters) programme is at the core of the 3P approach. It intends to transform Singapore's reservoirs and water bodies into clean streams, rivers and lakes with parks and gardens thereby creating new community and recreational spaces, and ultimately transforming the island into a 'City of Gardens and Water'. It represents a very comprehensive effort to make new community spaces attractive for people to enjoy activities close to the water and appreciate the value of the resource and of such amenities. Some 20 ABC projects have been implemented island-wide, and more have been planned for the near future.

Valuing water

In Singapore, there has been strong emphasis on 'valuing' water and thus on pricing it. The philosophy of pricing is based on water's rational use as explained in the PUB's vision document 'Water for All: Meeting Our Water Needs for the Next 50 Years' (PUB, 2011b). The rationale is that the next source of water could cost much more than current sources. This argument establishes the foundations of a realistic water-pricing regime that reflects the value of this resource to ensure its sustainability for future generations.

There are two components in how Singapore prices water. The water tariff part is aimed at covering the full cost of water production and supply. However, pricing is kept at a higher rate than just at cost recovery level as this may still result in water over-use (PUB, 2011a). The second part is comprised of a government-accruing water conservation tax, levied to the water tariff to reflect its real cost and thus the higher cost of developing additional sources of water. Nevertheless, valuing clean water involves much more than paying a price for its provision. For present-day Singapore, valuing water is about appreciating its fundamental importance for development, security, environment conservation and overall quality of life.

Further thoughts

With pragmatism, Singapore has designed and treaded its own development path by making the population aware of the importance of conserving water through different means. As expected, the overall communication, information and education strategies have been intertwined with the country's socio-economic and political course.

Apart from political crisis at the time of independence, the most vital immediate problems for the newly formed nation included the need to promote economic growth and social cohesion. Accordingly, from 1965 to 1972, as the government concentrated primarily on the growth of the economy, there was hardly any public information campaign on water conservation. During this period, the PUB focused on providing services to the entire population with efficiency and effectiveness. Water-related campaigns started in 1972 and, together with the 1968 Environmental Public Health Act, provided the framework for the public to appreciate the value of living in a clean environment.

By 1972, the 'economic miracle' had been achieved, about 80 per cent of the population were literate, and more than one-quarter of population were enrolled as students. Politically, the country was stable and people were ready to quickly respond to practical programmes launched by the government. At the same time, Singapore was almost totally dependent on imported water, with economic and population growth both accelerating water consumption. Elder citizens, who constituted the majority of the still illiterate population, were very supportive of the government messages since they had experienced water rationing in the early 1960s. Therefore, even though there was not much interaction between the government and the population, simple messages addressing the importance of saving water can be considered as the best possible demand management strategy.

Even though there are no empirical studies available on the impact of these campaigns on water consumption, looking at the water conservation pattern in the domestic sector in the 1970s and 1980s, one has to infer, if not to acknowledge, the positive impacts such initiatives indeed had. More importantly, these outreach actions made the majority of the population aware of the real need to save water. Direct evidence of the impact of non-interactive campaigns was seen during the unusually long dry period in 1990, when, as a result of extensive publicity programmes based on 'scare tactics', water consumption in Singapore dropped significantly.

Water education programmes have been implemented in schools under the philosophy that attitudinal change takes years. As the PUB officials mention, it takes two generations to make people aware of any specific issue and two more to entrench the message. There was thus a sustained focus on educating students on water-saving habits from the early 1970s.

By 1992, the literacy rate in Singapore was above 90 per cent and universal literacy among the 15–24 age group. By then, more than 80 per cent of the population was living in government-built high rises and per capita income was among the highest in the world. An affluent, well-educated and ambitious population had witnessed the transformation of the island into a clean, efficient first world city. The government was soon to acknowledge that among this generation were highly talented and professionally qualified individuals from whom ideas could be tapped through wider public consultation.

In the late 1990s, the government encouraged civic bodies to participate in public engagement activities in regard to water conservation, emphasizing active citizenry to promote socio-cultural and environmental issues. Nevertheless, the response did not reach the expected level. One of the possible reasons may have been the distinct civil society organization dynamics in Singapore compared to other parts of the world. Generally, in other parts of the world, civil society or non-governmental organizations adopt projects that are either not undertaken by any agency or where public sector efforts were less successful. In contrast to this, civic groups in Singapore were encouraged to be complementary to what the government was already doing successfully. Social groups were expected to activate the citizenry and motivate them to partner with government agencies and join their efforts. This was based on the principle of reinforcing the 'my Singapore' ethos among the residents.

A paradigm shift in the way water engagement campaigns were designed came with NEWater. At that time, the PUB realized that old slogans and non-interactive approaches in public engagement campaigns had become antiquated and stale. As the public no longer perceived water supply to be the main problem, campaigns had to be modified. Water management demand was thus framed by the notion of the Four National Taps. The new message was to be communicated to people in more subtle, attractive and impressive ways. Therefore, conservation campaigns evolved to their present scope and concepts like 'bonding with water', 'valuing water' and 'enjoying water' were introduced. These finally culminated in the ABC Waters programme where water's recreational value has been made universal with the development of attractive waterscapes all around the city-state.

Since 2004, the PUB has taken this partnership with the general public to a higher level. The new age strategy has been to make people aware of the importance of water quality through interactive public engagement. The ABC Waters programme was considered ideal to encourage people to get closer to water. It was deemed that, with availability of and access to attractive waterscapes all around, the value of water would not go unnoticed irrespective of whether people were at home, at work or at play. Waterfront activities would not be a privilege for affluent citizens, and the population at large could also enjoy scenic views from their housing units and join water activities at their doorstep. As a strategy, people have been encouraged to get close to water and take responsible ownership of it.

Nonetheless, more emphasis on partnerships and the creation of a personal citizen–water relationship does not mean that previous information and awareness campaigns are not taken into consideration. In fact, they have been pursued with more dynamism, keeping in mind the generational change. The fundamental belief is that the new generations of Singaporeans have never experienced water scarcity and thus they are different from older cohorts who insisted on savings and avoided wasting water. These campaigns have been also an attempt to bring the new generations on board and make them realize that even if water supply may not be the big question at present, its sustainable use is. As a result, different means are used to spread information and raise awareness among young people.

In spite of the value of education, information and awareness campaigns, there is no empirical data available that allow us to assess their direct impacts in terms of water conservation. Even though the PUB conducts perception or readership surveys once every two years or so, the focus is not on impact of the campaigns alone, but on the whole spectrum of the PUB activities. Some examples include evaluation studies conducted by Nielsen in November 2007, which revealed that 94 per cent of respondents felt that they had a part to play in Singapore's water management, an increase from 88 per cent in a similar study conducted in 2005. Another example is a PUB study focused on the acceptance of NEWater and which showed highly positive public response. As mentioned earlier, one can just assume that campaigns have been an effective instrument to create awareness on the importance of water conservation.

Finally, Singapore's long-term, comprehensive water resources strategy (policy-making, planning, management, governance and development) has been thoughtfully elaborated keeping in mind the fundamental part education, information and awareness play. It is well known that the implementation of plans, programmes and projects depends on their popular acceptance and support, and it is thus for policy-makers and politicians to provide means for the people to get engaged and become responsible for the management of their resources. In the city-state, cooperation, coordination and communication systems between different partners have been set and, more importantly, have been sustained along the years in spite of their complexity. Every effort to include people in the gears of the system, an undoubtedly more challenging task, can only be positive not only for the present generations but also for those to come.

6 Cleaning of the Singapore River and Kallang Basin

Introduction

While much attention in recent years has been paid to water scarcity issues, the problem that is likely to precipitate a major crisis in the foreseeable future is the deterioration of water quality. The issue remains neglected in most parts of the world, especially in developing countries, where rivers and lakes near urban centres are already grossly contaminated with known and unknown pollutants. It is indeed a development paradox that people all over the globe are clamouring for a better quality of life, and yet their economic, health and environmental conditions are not improving as per their expectations.

One of the main activities Singapore has implemented as part of its overall strategy for sustainable development has been the cleaning-up of its several river systems, many of which were significantly polluted when it became independent. Fortunately, the city-state had a visionary and dynamic political leadership which, by any account, was well ahead of its time. In 1969, the then Prime Minister Lee Kuan Yew asked the relevant government departments to draw up and implement a plan to clean-up Singapore's rivers. However, these achieved limited results as they were mainly concerned with the plan's implementation costs. In 1977, disappointed with the little progress attained over the course of eight years, the Prime Minister gave governmental bodies ten years to complete the operations. True to Singapore style, the strategy to clean up both the Singapore River (historically the island's most important trade artery) and the Kallang Basin was implemented within this timeframe. Prime Minister Lee's vision of Singapore can be clearly distilled from a cursory analysis of the positive impacts accrued to the restoration of these water bodies, improvements that include significant health, social and environmental benefits to the general population as well as the phenomenal increase in commercial activities and land values around the riverbanks.

This chapter analyzes the planning formulated by the government of Singapore to clean the Singapore River and the Kallang Basin and how it was successfully implemented. It also discusses the human and environmental dimensions of these strategies not only to improve the conditions of the river and its surroundings, but also to develop the city-state, and provide its population with an improved quality of life and a cleaner environment.

Historical background

Founded in 1819, Singapore became one of the world's most important and busiest ports by the late nineteenth century. With the development of steamships (1840s) and the opening of the Suez Canal (1869) the volume of trade soared and the economy flourished. By 1954, over five million tonnes of imports and exports commercially valued at $1.2 billion were conveyed along the Singapore River each year. Given its importance, this south-flowing stream came to constitute the island's main trade artery and most of the city developed outwards from its banks in the central area. For decades, the Singapore River and its banks were synonymous to the island. In fact, from the foundation of the colonial port city in 1819 and until the demise of the last *twakow* (a lighter craft or goods boat of Chinese origin used along the river) in 1983, the river gradually became the focal point of all global and regional trade passing through the island (Dobbs, 2003).

Flourishing trade brought more wealth to Singapore along with new problems, mainly with respect to water, its use and its quality. Increasing navigation started to constrain the river and growing activities along its banks attracted more population, squatter colonies, hawkers and backyard industries, including those processing gambier, sago and seaweed. The domestic and industrial wastewater and solid waste all of these settlers and economic activities produced was discharged untreated into the river, contributing to its increasingly severe levels of pollution (Tan *et al.*, 2009). Clearly, decades of rapid development and lack of long-term planning had resulted in overcrowding in the central area and the serious pollution of the river.

Watercourses

The Singapore River has a maximum navigable length of 2.95 km from the point where it starts at Kim Seng Bridge to its mouth at Marina Basin. It has a width that varies from 160 m at Boat Quay to 20 m at Kim Seng Bridge. The Kallang River, the longest in Singapore, flows for 10 km from the Lower Peirce Reservoir to the coast at Nicoll Highway. The Kallang Basin drains five main rivers: Bukit Timah/Rochor, Sungei Whampoa, Sungei Kallang, Pelton and Geylang. For its part, the Singapore River drains a catchment area of about 1,500 ha, whereas the total drainage surface of Rochor, Kallang and Geylang is 7,800 ha (Yap, 1986). It also joins with the five rivers from the Kallang Basin at the city front, and together they flow out to the sea through Marina Bay. Although they are not large rivers, their importance and relevance to Singapore is akin to that of the world's great rivers to the countries through which they flow.

Increasing trade and the related urban and industrial growth that occurred from the nineteenth century onwards very soon contributed to the heavy contamination of the Singapore River. As early as 1822, Raffles established a committee to look into the state of the river and found that there was a considerable amount of sand around its mouth because of the construction of jetties on the North Boat Quay side. During the 1850s, a dredger was built to remove sediment in the river, but the measure did not prove to be very effective and, by 1877, navigation had become extremely difficult as a result of silting. A government committee was then set up to further investigate ways of cleaning and deepening the river and a decision was made to purchase a dredger to scour the riverbed (Dobbs, 2003). From the 1880s onwards, river dredging became a regular activity but it could not solve the pollution problems that only worsened with time. Dobbs (2003: 3) draws attention to a report of the Singapore Municipality issued in 1896 stating that 'the outbreaks of cholera and diarrhoea in the settlement could be directly attributed to the insanitary state of the Singapore River'.

In 1898, the Singapore River Commission was established to solve the river's pollution-related problems. This new body highlighted the extraordinary number of vessels that transported goods along the river and recommended several alternatives to improve its conditions. Measures such as widening of the river, improving embankments, regular dredging, raising all bridges, and building a boat harbour either in the river or the sea near the river mouth were suggested in the report prepared by the Commission. These recommendations, however, were not implemented mostly due to financial constraints, as they involved land acquisitions and replacement or construction of bridges across the river. Subsequent recommendations to improve the conditions of the Singapore River made in 1905 and 1919 were also not carried out due to financial considerations.

The river was considered vital for the economy of Singapore and both the colonial government and private sector groups agreed that it needed to be widened and cleaned. Nevertheless, these tasks were too big in scope and required massive investments and significant political will. Such considerations continued delaying the decision to put into place the set of measures that had been suggested to clean the water body at different times.

In 1953, a new body, the Singapore River Working Committee, was established to control, maintain and improve the Singapore River as a trade artery within the harbour limits (National Archives, 1953). In September 1955, the Committee submitted its final report to the government (National Archives, 1955) with a series of recommendations to improve the conditions along the river and in surrounding areas to foster trade, as well as to control the chronic problems brought about by pollution. The cost of putting these recommendations in place was estimated at around $15 million (Dobbs, 2003). This was the last colonial attempt to find a road map to improve the river, but one that was also never implemented.

Subsequent Labour Front and PAP governments also emphasized the river's commercial and social importance. New recommendations were proposed, but similarly to all previous initiatives, no action was to be taken. Financial constraints stalled all efforts, as well as the river's importance as a major trading artery where activities could not be disrupted. And yet, with the passage of time, the Singapore River was only becoming more polluted with rubbish perennially floating in its waters (National Archives, 1967).

The extent of contamination was such that the stench extended all throughout the area surrounding the watercourses. Discharges of organic and inorganic waste from pig and duck farms, squatters, backyard trades, cottage industries, latrines, street hawkers, vegetable vendors, boat repairers and similar riverside activities, sullage water and refuse from unsewered houses were all poured directly into the streams. These had made not only the Singapore River but also all rivers in the city-state grossly polluted and with no signs of any aquatic life (Ministry of the Environment, 1987).

Planning for development

One of the most pressing problems post-colonial Singapore was to face was the serious housing shortage. The inclusion of this serious issue in the priority list of the PAP's election campaign was swiftly followed by specific actions once the party won the general elections in 1959. Upon self-government, planners noted that:

> Occupancy figures in many of the shop-houses in the Central Area had soared to over 20 times the original one family occupancy by successive sub-divisions, with residential densities reaching over 2500 persons to the ha. In addition, there was extensive squatting on vacant and marginal lands within the Centre, with people crowding in huts composed of largely inflammable material and devoid of sanitation, water or any of the elementary public health requirements.
>
> (Tan, 1972: 334–335)

In 1960, it was estimated that approximately 250,000 people, in addition to nearly 250,000 squatters, were living in city slums. It was calculated that no less than 147,000 new homes would have to be built during the 1961–1970 period in order to solve the city-state housing problem.

The ruling PAP set a five-year plan to improve the life of the population, to give citizens a sense of ownership, and to address the many challenges related to the economy, the lack of housing and overcrowding in the central area. The first and main piece of legislation to this effect was the Housing and Development Act of 1960, which gave the Minister of National Development the authority to put in place a low-cost housing programme (Quah, 1983). This initiative included the clearance of slums as well as the construction, conversion, improvement and extension of any building for sale, lease, rental or other purposes. To enact such policies, institutions were created to develop public housing units and promote economic growth. Starting operations from 1 February 1960, the Housing & Development Board (HDB) became the first statutory board, reflecting the priority attentions that had to be allocated to tackling the serious housing shortages. This was followed by the creation of the People's Association (PA) on 1 July 1960 to deal with the communist and communal threats by controlling and coordinating the 28 community centres it was to oversee. One and a half years later, the Economic Development Board (EDB) was established to solve the island's growing unemployment, a challenge that was to be faced by attracting foreign investment to Singapore (Quah, 2010).

Immediately after Singapore achieved independence in 1965, its overall rate of economic growth was unprecedented. GDP increased by a compounded annual rate of over 9 per cent during the 1960s and the industrial production increased by more than 20 per cent. For instance, the output contributed by the island's shipbuilding and repairing business almost doubled from $64 million in 1966 to $120 million in 1968. That year, Singapore was made the headquarters of the Asian Dollar Market, and one year later, in 1969, it became a gold market so large that it surpassed those of Hong Kong and Beirut and a port so important that it became the busiest in the Commonwealth, surpassing London. Moreover, after the town of Jurong was reorganized into the Jurong Town Corporation, manufacturing activities soared in the 11 other industrial estates it managed. By the end of 1970, Jurong housed 264 production factories employing 32,000 workers and another 106 factories were under construction (Turnbull, 1977). By 1970, near full employment levels were recorded and the city-state was confronted with the need to relax its immigration laws (Turnbull, 2009).

Throughout the 1960s and under the PAP government, Singapore witnessed yet another unprecedented change, but this time regarding urban renewal. Dilapidated sections of the city were redeveloped and slums were cleared and replaced with modern high-rise buildings. During this decade, thousands of public housing units were constructed by the HDB and much of Singapore's population was moved to the new towns recently developed outside the central city district. From then, the government asserted that this growth should not be at the expense of the natural environment. On the contrary, it was recognized that a clean and green environment was essential to attract investment and retain talent so essential to support further growth. The city-state thus invested in creating critical environmental infrastructure right after the independence, despite competing demands for funding. For example, and among many supported initiatives, $2,000 million was spent on drainage development projects, $3,600 million was earmarked for the first phase of the Deep Tunnel Sewerage System, and $300 million was allocated to clean the Singapore River throughout the years (this figure varies with the sources; for example, Chou (1998) mentions a smaller figure of $200 million).

Change triggered by political will

Urban reforms were essential for Singapore's development and to significantly bring down pollution in the watercourses. This was a challenge the then Prime Minister Lee Kuan Yew did not conceive as an isolated problem. He rightly observed that river pollution was the end result of all other pollution problems prevalent in the city-state. If the nation would develop as a productive industrial society and the population would be offered an improved quality of life, the rationale was to envisage, plan and implement solutions to the issues affecting the general population, and those living on the periphery of the Singapore River in particular.

In March 1969, Lee called on the drainage engineers in the Public Works Department and water engineers in the Public Utilities Board (PUB) to work together on a plan to solve the environmental problems associated with the waterways. He stressed the necessity to control river pollutants and emphasized the restoration of the riverbanks. In his memo to the Head of the Public Works Department, Lee Kuan Yew clearly outlined his goals to 'keep these rivers and canals with more or less clean translucent water, where fish, water lilies, and other water plants can grow' (National Archives of Singapore, 1969a). A flurry of activities took place as part of this initiative. Within two weeks, the main agencies held meetings, pollution sources were identified and solutions to reduce the river pollution levels were proposed. Unlike pre-independence administrations, the Prime Minister was interested in more than just discussions and requested direct notifications on how the initiatives were proceeding. For example, in his letter to the PUB and other departments, the Director of Public Works, Lionel de Rosario, reinstated that feedback had been requested from the Prime Minister (National Archives of Singapore, 1969b).

In fact, the Prime Minister had already expressed his interest in cleaning the Singapore River even before the aforementioned memo. In 1968, his government passed a strict Environmental Health Act which allowed for the prosecution of anyone found to be polluting the rivers and waterways. At that time, the discussions and progress reports regularly referred back to the Prime Minister and a sense of urgency and expectation for rapid results was reflected in the activities undertaken (National Archives of Singapore, 1969c). An Atmospheric and Effluent Pollution Study Group was formed to carry out detailed analyzes of the riverbed, water and air and determine what pollutants and gases were present in and around the river. As expected, the study concluded that the river was highly contaminated with organic matter. As a technical solution, several tonnes of sodium hypochlorite were spread at specific locations to mitigate this situation, but it was quickly realized that this was not a practical way to tackle pollution (National Archives of Singapore, 1969d). The interim reports produced during 1969 identified solid and liquid domestic and industrial waste as the main sources of pollution (National Archives of Singapore, 1969e). The reasons behind these sources were also listed. For example, much of the domestic waste came from people living along the rivers or in the river catchment areas, with old settlements like Chinatown representing a source of considerable water pollution. It was then decided that hawkers, squatters, labourers in makeshift industries (with the exception of the lighterage industry), employees at storehouses and all others who made their living alongside the river would have to be relocated to other areas as soon as possible.

In August 1969, boat builders and firewood and charcoal dealers along Pulao Saigon Creek were among the first groups to be informed that they would have to move away from the river, as they constituted significant sources of pollution. All along the river, notices were served to businesses and individual premises. Provisions were made through HDB for affected individuals and businesses to receive priority in the allocation of housing units and premises for commercial use (Dobbs, 2003; National Archives of Singapore, 1970).

Relocation campaigns received broad media coverage. Nevertheless, the groups to be resettled were reluctant to move even after receiving compensation offers and new and much improved housing and/or businesses facilities. Some MPs expressed their concern over the actions that would be taken against unauthorized facilities established along the embankment of the Singapore River (National Archives of Singapore, 1971a). Since the government was very keen on carrying out the operation with as minimal confrontation and bad press as possible, the resettlement of unauthorized hawkers (National Archives of Singapore, 1971b) and squatters continued, but the speed of the removal proceedings remained slow. Overall, between 1969 and 1979, some 3,959 people of squatter communities were relocated (Tan, 1986; Dobbs, 2003). In 1971, the hawker settlement programme was introduced and approximately 5,000 hawkers were relocated to hawker centres (Tan *et al.*, 2009).

The cleaning operation: 1977–1986

By 1977, much of the environmental work and control activities of river polluting sources were already planned or under consideration by the various authorities who were responsible for the rivers. The cleaning of the various waterways had progressed close to the mouth of the basin, but the mouth itself and the catchment areas still represented a major challenge before significant improvements in water quality could be ensured: more than 46,000 squatters were still living in unsanitary conditions in the vicinity of the rivers; liquid and solid wastes from the hawkers, vegetable vendors, markets and unsewered premises all continued to represent a problem; and 610 pig farms and 500 duck farms were still draining untreated waste into the waterways, especially into the Kallang Basin (Dobbs, 2003).

Despite HDB's fast pace in constructing new flats, by 1977 there were still 46,187 premises with squatters, the majority of which used nightsoil buckets or pit latrines. Some had overhanging latrines that discharged waste directly into streams and rivers (Chou, 1998). The Kallang Basin had the largest number of squatters with 42,228, and the Singapore River catchment with 3,959. Moreover, the waste generated by the 4,926 hawkers and wholesale vegetable vendors who worked along the roads mostly at Upper Circular Road ended up putrefying in the drains and eventually polluting the rivers. The industries on the banks of the rivers, such as trading, lighterage, cargo handling, boat building and repairing were housed in old and congested buildings. Due to the absence of pollution control facilities, oil, sullage water and solid wastes were discharged to the river, worsening an already worrying situation.

Meanwhile, Singapore River's importance in terms of entrepôt trade changed. By the 1970s, the Port of Singapore Authority (PSA), which operated five maritime gateways (Keppel Wharves, Jurong Port, Sembawang Wharves, Tanjong Pagar Container Terminal and Pasir Panjang Wharves), was handling most of the cargo moving through the island. That decade, port facilities were exponentially developed, making Singapore one of the busiest harbours in the world. Technological improvements in cargo handling played a major role in this rapid transformation. By mid-1972, Singapore was already opening the first container port in the region, with two more under construction and due for completion in 1978. These actions led to the precipitated deterioration and soon insignificance of the Singapore River's vital role in terms of trade, as it rapidly became a 'graveyard for derelict lighters' (Dobbs, 2003: 110). Once it was clear that the river no longer played the key role it once had in Singapore's trade exchange, it became easier for the government to earnestly proceed with the cleaning operations.

On 27 February 1977, during the opening ceremony of the Upper Peirce Reservoir, Prime Minister Lee Kuan Yew gave a definite time window and deadline to the Ministry of the Environment (ENV) to clean the Singapore River and the Kallang Basin. His message was to make of keeping water clean a way of life, adding a future goal of '[i]n ten years time [...] have fishing in the Singapore River and fishing in the Kallang River. It can be done'. He further elaborated his vision by stating that '[b]ecause in ten years the whole area would have been developed, all sewage water will go into the sewers and the run-off must be clean'. He warned that anyone who failed to perform satisfactorily or placed obstructions to achieve this goal 'would have to answer for it' (Hon, 1990: 41).

The urgency and seriousness shown by the Prime Minister was clear to every government agency. This was evident from the fact that the fund requests sent by the ENV to lead the proposed tasks to clean the rivers were agreed to immediately by the Finance Ministry. Following the stipulations made in the Draft Plan, in areas where pollution control facilities had been provided, it was to be ensured that these were used and operated efficiently. In those areas where facilities had not been provided, but were possible to be put in place, urban development targets would be set. Finally, in the remaining areas where it was either impossible or not feasible economically to provide facilities, for example, roadside hawkers, boat colonies, etc., pollution sources would be controlled, minimized or eliminated (Hon, 1990).

A thorough field survey to identify the various sources of pollution was initiated in June 1977 and completed in September 1978 (Chiang, 1986). Three types of human pollution sources were identified. Type I sources included areas where houses, hawker sites or factories were still not connected to sewers. The solution in this case was to provide sewerage facilities and educate householders and health service staff to use the appropriate means to dispose of waste. Type II sources comprised pollution in unsewered land (mostly from squatters' bucket latrines) to be later developed, an issue that could be ameliorated by the prompt resettlement of squatters. Type III sources had to do with areas where pollution-control facilities were not available and there were no definite plans to redevelop the area or where the provision of such facilities was not feasible. Such sections included unproductive pockets of land with squatters, boatyards and pig farms, all of which needed to be vacated (Hon, 1990).

Under the Permanent Secretary, ENV, a Steering Committee was established to look into the Draft Plan that ultimately led to the preparation of a Master Plan. Once approved by the Cabinet, the document became a government edict to be implemented by the relevant government agencies under the coordination of the ENV (Ministry of the Environment, 1981). The Plan acknowledged and pointed at the complexity of the pollution problem and the need to involve the various ministries and government agencies that had a role to play in the cleaning process. Apart from the ENV, some of the other ministries that would be involved directly and indirectly in the project included the Ministry of National Development (MND), Ministry of Trade and Industry (MTI), Ministry of Communication & Information (MICA), Urban Redevelopment Authority (URA), JTC, Primary Production Department (PPD), PSA, Public Works Department (PWD) and Parks and Recreation Department. To avoid interagency conflicts, it was deemed necessary to integrate the entire project to the nation's long-term strategic development plan so that it was not carried out in isolation as had happened several times before.

As the catchments represented approximately 30 per cent of Singapore's area, planners were confronted with the challenge to list a set of proposals on how to prevent polluting activities of very varied nature and that were also located far from the rivers (Tan *et al.*, 2009). Therefore, an integrated management action plan was devised addressing not only the immediate steps necessary to clean the rivers but also the long-term approach to be followed to maintain them pollution-free (Chou, 1998).

By the time the programme to clean the rivers started in 1978, some 21,002 unsewered premises had already been identified in areas like Geylang, Kim Chuan, Kg Bugis, Jalan Eunos, Havelock Road and others (Chiang, 1986). These areas were densely populated with squatters and the difficulty to access them made the cleaning operation almost impossible. Most of these unsewered premises were served by nightsoil buckets, pit and overhanging latrines, which were unsanitary, a source of unpleasant odour and pollution, and presented a health hazard in general. At the beginning of the programme, 11,847 nightsoil bucket latrines were identified, all of which were phased out except for 533, which were sewered. At last, the last nightsoil bucket was phased out in 1987 as well as 621 overhanging latrines. Additionally, in 1977 the 3,961 unsewered premises discharging sullage water into the watercourses were reduced to 36, and out of 710 premises without refuse-removal services, only 129 remained by September 1981 (National Archives of Singapore, 1981). What is more, sand-washing was controlled by phasing out private sand quarries in the areas of Tampines and Bedok and by centralizing such activities under a public holding company.

The action plan's main goal was to restore the Singapore River and the Kallang Basin to the extent that aquatic life could thrive in them. This was expected to be achieved through five major action lines: the removal or relocation of polluting sources and phasing out of polluting industries; the development of suitable infrastructure for those affected by relocation; raising awareness about the overall development programme; strict law enforcement; and by cleaning and dredging the waterway.

The following sections detail some of the many challenges faced both by the government and the general population during the implementation of the cleaning operation.

Resettlement, relocation and phasing-out activities

The Resettlement Department, under the HDB, was the main authority overseeing land clearance and the resettlement of squatters. As part of the cleaning up programme, high priority was given to the clearance of squatter colonies and rundown urban areas in order to proceed with redevelopment. An eight-year target was established for the completion of this programme (Tan, 1986).

Squatters were resettled under a Resettlement Policy first introduced during the 1960s. A measure that only applied to Singapore nationals, all persons and business establishments affected by resettlement were to be offered housing and compensation (Tan *et al.*, 2009). In 1978, about half of 46,187 squatters were on privately-owned land. According to the resettlement plan, the land-acquisition authorities would first acquire the land before it could be cleared. Furthermore, the authorities also had to respond to the major challenge of constructing suitable infrastructure in new areas to relocate a large number of shops and backyard industries.

The Resettlement Department classified the squatters into three categories based on the amount and type of pollution they generated. Serious sources of pollution were urgently cleared for which some 675 pig farms were scheduled for complete clearance by 1981 as a top priority. The second most pressing matter had to do with squatter colonies in the catchment areas that were to be relocated in housing estates in new towns. Finally, the pockets of land that would become open spaces, greenery or used for future developments, were left for the later stages of the clearance process.

To conduct the resettlement exercise in an orderly manner, clearance and relocation programmes were based on census surveys of the affected squatters. Depending on who owned the land, eviction notices were given by the Land Office, Resettlement Department, HDB and JTC. Sometimes several notices were handed out, including repeated warnings to convince squatters to relocate. Severe actions were then taken against non-compliant squatters as well as against those delaying their departure without valid reasons.

In general, all persons and business establishments affected by resettlement were offered monetary compensation as well as housing units that were of much better quality than the ones they had before. They were also granted rent concessions or waivers in the downpayments to acquire their new facilities. Squatters were paid compensation on fixed government approved rates, which was ex-gratia in nature. For example, farmers were paid approximately $205/m^2 for their house, while squatters were offered $105/m^2 of housing. Also, cash grants were given to farmers in lieu of alternative farm lands, and residential families and business establishments living in the central urban areas that preferred to arrange for their own accommodation were also paid cash compensations (Tan, 1986). Table 6.1 shows the number of squatters cleared both in the Singapore River and the Kallang Basin.

Seeking to minimize the inconveniences related to resettlement from a familiar to a fairly new environment, a wide range of benefits was extended to the affected population. At the same time, political stability and fast economic growth during that period also helped make resettlement a less painful process.

Table 6.1 Squatters cleared in the Singapore River and Kallang Basin

Catchment area	Target number of squatters to be cleared	By 1979		By 1981		By 1983		By 1985 (September)	
		No. of squatters cleared	%	No. of squatters cleared	%	No. of squatters cleared	%	No. of squatters cleared	%
Singapore River	3,959	1,097	27.7	1,921	48.5	3,213	81.2	3,744	94.5
Kallang Basin	42,228	9,657	22.9	24,781	58.7	34,596	86.9	40,830	96.7
Total	46,187	10,754	23.3	26,702	57.8	37,809	81.9	44,574	96.5

Source: Tan (1986).

Close and effective coordination between the Resettlement and Planning departments guaranteed that alternative housing was constructed before the squatters were moved. Nonetheless, resettlement proved difficult for businesses as most people continued to continuously reject the sites proposed for their new facilities but eventually communities and businesses settled down in the new places offered. This, in turn, proved beneficial for the initially reluctant population in the long run (Hon, 1990). For example, Kampong Bugis, located within the catchment of the Kallang Basin, was considered to be a fire hazard, a death trap and a slum close to the city, in addition to being one of the main sources of pollution for the basin. In 1982, the area was inhabited by 1,137 families totalling 3,000 people living in wooden or zinc houses, squeezed into every inch of land and without proper sanitation. There were also around 151 industries of different sorts, five large warehouses, 12 boatyards, 50 shops and 12 Chinese temples. Now, after two decades of the river clean-up, this area is among one of the most urbanized landscapes in Singapore (Mak, 1986).

By 1986, all squatters had been resettled, meeting the eight-year deadline that had been set to clear the catchment area. The emptied areas were then developed by the HDB and trasformed into housing estates, new towns and industrial parks; e.g., Kampong Bugis, Bishan New Town, Potong Pasir, Serangoon New Town, Eunos Estate, Kampong Ubi Industrial Estate, Geylang Estate, etc. Initially, resettlement was painstakingly slow mainly because of the shortage of alternative accommodation for the squatters and their shops, as well as industries. To expedite the process, the JTC requested support from the HDB, after which an announcement was made that more flats for relocation would be available in 1984 and 1985.

For the local industries, specific-purpose workshops were built at Lorong Tai Seng and Kampong Ubi while some others were also relocated in factories at Kallang Place in modular units of 100 m² rentable areas to suit their needs. Lastly, both boat builders and repairers of Kampong Bugis were relocated in suitable places at the Northern Tuas Basin in Jurong. Within two years of having identified the location, roads, drains, bridges and sewers had been completed and the resettlement process started. Finally, by December 1985, all of them had been relocated.

Street hawkers

In 1960s, with the nation-state's rapid industrial and economic development and accompanying abundant employment opportunities, more people joined the labour force. The demand for cheap and convenient food on the streets grew and thousands of people saw an economic opportunity in becoming hawkers. According to Tan *et al.* (2009), at one point in the 1960s, there was nearly one food seller for every 100 people. This created serious problems in terms of human and environmental health as the lack of solid waste management resulted in food waste generated by street hawkers being indiscriminately thrown onto the public roads. Many of the street-hawking sites were in the Singapore River's catchment, which meant solid waste eventually ended up in the river (Kuan, 1988). The situation became so serious that, in 1968, no new hawker licences were issued, except for those cases where it was considered that the applicant really did not have any other option to earn a living.

In December 1978, the number of street hawkers in the catchment area of the Singapore River and the Kallang Basin reached 4,926; most of them concentrated in Tew Chew and China streets in the Chinatown area (Chiang, 1986). At peak hours, it was impossible for any vehicle to circulate around these areas due to the presence of more than 1,000 hawkers (Hon, 1990). Apart from these street food vendors, vegetable wholesalers also conducted their business activities in the area, most of them traditionally operating in narrow footpaths, streets and vacant land without proper facilities. Large amounts of discarded vegetables usually ended up putrefying in the roadside drains.

The government embarked on a programme to build food centres and markets to which licensed hawkers could be transferred and moved away from the streets. Initially, food sellers did not like the idea of relocating to new places. To encourage them, officials from ENV held long discussions with their representatives and with the Citizens' Consultative Committees (CCC) of the area. This relocation also brought disagreements among different government departments. For example, at Bugis Street, the Singapore Tourist Promotion Board did not approve of the idea of moving the street hawkers elsewhere since it considered this measure would render the place unattractive. The Tourism Board thus insisted that hawkers were relocated to places that would retain the same ambience and with similar street scenes resembling their original conditions. These demands were disregarded when, after the removal of the street hawkers, all that was left in the places they vacated was an unsanitary place (Hon, 1990). By December 1985, all street hawkers had been relocated to either markets or food centres (see Table 6.2).

Table 6.2 Resettlement of hawkers situated in Singapore River and Kallang Basin areas

	December 1978			December 1985		
	Singapore River area	*Kallang Basin area*	*Total*	*Singapore River area*	*Kallang Basin area*	*Total*
No. of street hawkers	2,421	2,505	4,926	79	0	79
No. of street hawking sites	90	169	259	2	0	2
No. of hawkers in market/food centres	4,960	8,026	12,986	5,851	9,217	15,068
No. of markets/food centres	43	58	101	47	65	112

Source: Chiang (1986).

The workload of disposing of rubbish after each relocation proved to be a massive exercise. For example, cleaning up Chinatown took three months and the refuse removed amounted to 390 tonnes. A massive rodent control operation was later launched to eliminate rats and other pests in the area (Teo, 1986). Alongside these operations, the Environmental Health Department (EHD) carried out an extensive health education programme that included the distribution of health education materials, awareness-raising activities among school children, interaction with the residents and appeals for the public participation of community leaders.

Lighterage industry

In the Singapore River and the Kallang Basin catchments, the main riverine activities included trading, transport and boat building and repairing, which were associated with warehouses and/or workshops located along the river banks. There were thus a number of factors behind high levels of river pollution, for instance the inadequate sewerage facilities in the area, sullage water, garbage and waste from boat operators and their families whenever they lived in the boats, as well as from diesel and engine oil leakages. Therefore, the relocation of riverine activities became essential in order to achieve the clean-up of the rivers. The ENV noted that 'there were many sources of pollution in the catchments of both rivers – one of the main sources being lighterage and related activities. The only practical and economic solution is to resite the lighters' (*The Straits Times*, 9 August 1980: 9).

The responsibility of implementing this programme component was given to the PSA and JTC. The task to relocate the lighterage industry from the traditional base it had occupied for more than 150 years to a completely new area at Pasir Panjang proved a rather difficult one. A comprehensive relocation study was conducted and new facilities were provided to the entire riverine community at an expense of approximately $25 million, excluding land costs. New facilities were developed in consultation with the lighterage community through the Singapore Lighter Owners Association (SLOA) and, whenever possible, its recommendations were taken into consideration to support lightermen in view of their long history associated with the river. This did not stop the lighterage industry's vigorous campaign of 1977–1983 to be allowed to continue working along the river, even arguing that its actions were not responsible for the pollution (Poon, 1986).

At Pasir Panjang, the PSA provided 19 additional lighter berths to handle cargoes. It also commissioned 100 m breakwaters to shelter the wooden lighters from the onslaught of the south-west monsoon as a response to the lightermen's complaints about the lack of proper anchorage at this new site. Their objections were based on arguments that the waves were stronger at Pasir Panjang compared to the sheltered water of the rivers, and also that the place was too far away for the majority of lighter operators who lived in the Chinatown area (Tan *et al.*, 2009). Additionally, lighter repair facilities were provided at Terumbu Retan Laut and about 160 buoys were installed for mooring the lighters. A jetty was also commissioned to help lightermen commute easily between the mainland and the mooring area and improved facilities such as offices, rest houses, toilets, canteens, etc. were provided for the workers.

The relocation of some 800 lighters was carried out in two stages. The first one was completed in September 1982, when all the lighters from Marina Bay, Telok Ayer Basin, and Rochor and Kallang rivers moved to Pasir Panjang. In 1983, the rest of the lightermen from Kallang, Singapore and Geylang rivers were also relocated (Poon, 1986). By 1985, only six out of 64 boatyards remained in the Geylang River, all of which agreed to implement pollution control measures. By December 1986, charcoal-trading industries operating along the Geylang River were relocated to Lorong Halus, where the HDB undertook a $5.66 million investment to build proper facilities (Ministry of the Environment, 1987).

Relocating the lighterage industry as part of the clean-up operation was justified by the fact that this industry no longer played the key role it once had in Singapore's port commerce (Dobbs, 2003). In moving the craft boats, 120 tonnes of flotsam and rubbish were removed and disposed of from Crawford Street, under the spans of Merdeka Bridge and along the Rock Bund at the Kallang Basin. In the Singapore River, in August 1983, more than 260 tonnes of rubbish were disposed of at Boat Quay.

Pig farms

Pig farms flourished in Singapore after the Second World War (Hon, 1990). They soon became a serious source of pollution as they produced not only one of the highest amounts of solid waste, but also one of the highest concentrations. In 1971, during the development of the Kranji-Pandan Reservoir, an experiment was conducted to relocate traditional pig farms at Kranji to high-rise intensive pig farm units at Punggol, an area outside the water catchments at that time. Unfortunately, this did not solve the problem of pig waste management for several reasons. One of them was that processing the produced waste for disposal required large extensions of land that were simply not available in Singapore (Khoo, no date).

As such, the decision to phase out pig farming dates back to as early as 1974, when the first unprotected catchment reservoir in Kranji was built. However, the process was accelerated when the programme to clean the water reservoirs and rivers was launched. In 1974, about 10,000 local pig farms (with more than 716,500 pigs) around the island produced a substantial amount of fresh food to meet the domestic requirements. In the Kranji catchments alone, there were 2,926 pig farms.

Initially, large farms with more than 100 pigs were given the option to relocate to Punggol, an area then considered safe from the water catchment pollution point of view. Smaller pig farms were required to gradually phase out pig rearing and convert to non-polluting farming activities, such as poultry. Nonetheless, in March 1977, all the farms that had been allowed to remain in Kranji were asked to discontinue pig rearing completely. In October 1979, it was decided that pig farming should also be terminated from all water catchment areas in the city-state, including the Singapore River and Kallang Basin catchments, where there were more than 3,259 farms. A notice was issued to all the farms to stop their activities within two years and farmers were offered the option to be resettled and receive benefits provided they decided to engage in a different economic activity altogether. At the same time, in Punggol larger pig farms with the capacity to transform into intensive modern pig farms were given the option to apply for land permits at Punggol to develop such activities (Loy, 1986).

Out of the 3,259 pig farms located in these catchments area, 963 opted to receive resettlement benefits. The remaining 2,296 chose to remain in the same land and convert to non-polluting farming activities. By March 1982, all pig-rearing activities were phased out from these areas and no more pigs were reared in water catchments, including the Singapore River and Kallang Basin. At the same time, in Punggol 670 ha of land were habilitated with modern facilities for pig farming. In early 1984, and despite the fact that Punggol was not a water catchment area, serious environmental problems related to the management of large volumes of animal waste led to the government decision to completely phase out pig farming in Singapore (Loy, 1986).

Duck farms

In February 1980, more than two million ducks were reared in Singapore's 7,290 duck farms, many of them in the Kallang Basin. Ducks were reared in the open, polluting the surroundings as well as the water catchment areas. The extent of the problem was such that, in April 1980, the government decided to discontinue open range duck farming. The PPD started to notify all licensed farms to phase out this activity within one year but by April 1981, only about half of the farms had closed. Reminders were sent requesting farmers to phase out their farms immediately. They were also notified that no time extensions would be issued, except for valid reasons, such as in-progress construction of proper duck houses, and that the PPD would assist those interested in building appropriate facilities, which it did when requested. By December 1981, 80 per cent of open duck farms had closed. For the remaining 20 per cent, the ENV had to issue notices for summary prosecution for causing pollution. Finally, by March 1982, all farms rearing ducks in the open had been closed (Loy, 1986).

Public health, education and enforcement

As already mentioned in this chapter, the EHD carried out an extensive health education programme accompanying the cleaning operations (Teo, 1986). From 1968, campaigns to raise public awareness on environmental matters have been carried out both through the formal education curriculum and awareness events aimed at building public awareness. These have covered a wide range of issues including pollution, food hygiene, infectious diseases, waste management, sanitation, anti-spitting, anti-littering, river clean-up, and global environmental issues. Often, a campaign precedes the introduction of a environmental or public health law for the general public to become acquainted with the new provisions and their implications.

Since the habit of littering and the use of drains as a convenient channel of waste disposal was a big problem, the sources of pollution were checked regularly. During the river cleaning programme, the EHD provided the main labour force to clean up the public areas. A vigilant focus on surveillance and enforcement against incidental and accidental discharge of sullage water and refuse from markets and food centres, bin compounds, commercial premises, households and construction site washings was the key to instil a sense of discipline among the public and change more than 100-year-old habits.

From early 1979 to March 1982, when enforcement was at its peak, a total 11,409 inspections were carried out in sewered areas and further 4,139 in unsewered areas. During those scrutiny exercises, 148 persons were prosecuted and 712 were served with notices. To monitor the discharge of sullage water from domestic sewered premises 155,292 inspections were conducted, 1,179 warning letters/notices were served and 80 persons were prosecuted. In order to inculcate a culture of cleanliness among hawkers, 61,509 inspections were led to check eating establishments after which 107 polluting premises were prosecuted. Some 86,569 premises in markets and food centres and 94,051 bin centres were also inspected. During this period, 7,615 persons were prosecuted for littering (Ministry of the Environment, 1987). Later on, between August 1982 and 2005, more than 2,600 people were prosecuted for littering, dumping refuse and discharging sullage water in public places (Teo, 1986). Rigorous monitoring strengthened law enforcement and facilitated the cleaning operation and also served to encourage the population to adopt more sanitary and orderly habits.

Achievements and lessons learnt

The many and massive operations carried out to improve the island's environment, including cleaning Singapore's rivers, mainly the Singapore River and the Kallang Basin, involved making numerous changes and facing myriad challenges.

This process also produced innumerable positive outcomes, achievements and lessons learnt in terms of overall social and economic development as well as for environmental protection in the city-state.

Regarding provision of public housing, the HDB decided to erect housing estates within a radius of five miles from the city centre. This was because most of the people's activities were clustered in and around the central area and thus inhabitants were not prepared to move to housing estates in communities unfamiliar to them and away from their jobs if transportation costs were high (HDB, 1963; Waller, 2001).

Public housing thus provided an opportunity to relocate the population from slums to high-rise buildings and, in the process, upgrade the level of services provided in terms of water supply (from community standpipes to direct household water supply, gas and electricity). To give an idea of the extent of the public housing units that were necessary, it was estimated that the demolition of each shop-house in the central area would require the construction of a minimum of five to eight flats to relocate the affected population (Chew, 1973).

This large-scale urban development process called for the resettlement of 26,000 families, most of them moving into HDB-built public housing, with much improved access to and delivery of basic services, community amenities and transportation facilities. Consequently, these households saw a significant improvement in their quality of life. For the sake of public hygiene, all the 4,926 hawkers were relocated into the food centres the HDB, URA and ENV had built. This exercise was executed in such an efficient and scrupulous fashion that, by 1986, there were no unlicensed hawkers in Singapore.

Furthermore, the vegetable wholesalers previously operating at Upper Circular Road were located into the Pasir Panjang Wholesale Market in January 1984. These facilities were also built by the HDB at a cost of $27.6 million (Ministry of the Environment, 1987). More than 2,800 industrial cases of backyard trades and cottage industries were also relocated, most of them moving to those industrial estates built by the HDB and JTC. Finally, regarding farming activities, by March 1982, PPD had phased out all pig and duck farms from the catchment areas, significantly reducing pollution. What all of these exercises highlight was the public need and leadership willingness to tackle challenges and problems at the root, even when this meant long-term and apparently costly public investments, rather than opting for piecemeal policy decisions.

By September 1983, lighterage activities carried out by some 800 lighters were relocated to Pasir Panjang, where the PSA provided mooring and upgraded facilities at a cost of $25 million. The task of physically cleaning up the rivers became easier and speedier once the lighters no longer circulated along the river. From 1982 to 1984, 2,000 tonnes of refuse were removed from the Singapore, Kallang, Geylang and Rochor rivers, considerably improving the sanitary and aesthetic composition of the landscape in those areas (Poon, 1986). The Drainage Department dredged approximately 40,000 m^3 of sediments from the stretch of the Singapore River and about 600,000 m^3 from the Rochor and Kallang rivers (Yap, 1986). In December 1986, the charcoal trade was relocated from Geylang River to Lorong Halus where appropriate facilities were constructed by HDB at a cost of $5.66 million.

Relevant to this book's scope and the development of Singapore, the public housing programme had an enormous impact on water supply for both planners and users. On one hand, it required water planners to swiftly expand water supply networks and on the other hand, the scheme contributed to provision of piped water supply, an increase in the number of metered consumers and the consequent recording of water consumption patterns. As has been noted earlier, the number of HDB units rose tremendously from 19,879 in 1960 to 118,544 in 1970. Since each flat was provided with metered and direct piped water supply, the number of metered connections consequently rose from 102,819 in 1960 to 264,314 in 1970. Given this vertical growth in housing units, more than 90 per cent of the rise in installed meters came from the newly built public flats, once again drawing attention to the absolute necessity of agency coordination and comprehensive planning.

Further improvements were also made to the waterways. For instance, the length of the distribution and supply main increased from about 1,200 km and 80 km in 1960 to 1,840 km and 104 km respectively in 1970. More than 65 per cent of this expansion in the length of the distribution mains was to serve villages and HDB estates outside the city area. Thus, the HDB development programme played an important role not only in extending water supply coverage, but also in making sure that those consumers connected to the network were also metered. Notably, this remarkable improvement was made even before the creation of long-term plans such as the 1971 Concept Plan and the 1972 Water Master Plan. Thus, even without these strategies, institutional coordination between the HDB and PUB allowed the PUB to develop the necessary water supply infrastructure to ensure that the new housing developments would be completed successfully and on time.

Water quality measurements pointed at important improvements within the first three years of the programme to clean up the rivers. Sample analysis showed that the amount of organic pollutants had decreased considerably and the level of dissolved oxygen had gone from almost zero to 2–4 mg/l. Table 6.3 shows the improvement in the quality of river waters. Cleaner water quickly proved that restoring the rivers' flora and fauna, which was one of the main goals set for the cleaning-up programme, was possible.

Biological surveys carried out by the National University of Singapore (NUS) indicated that aquatic life, both pelagic and benthic, was returning to the waterways. A 1986 study showed that some 20–30 aquatic species had been found in the Singapore River and Kallang Basin (Ministry of the Environment, 1987).

Fish and prawns could already be caught from the rivers by 1983, well ahead of the set target, which prompted new legislation permitting only hook and line fishing (Chou, 1998). Water quality had improved to such an extent that towards the end of 1987, two dolphins were spotted in the Geylang River, where they stayed for a week. In September that year, the ENV celebrated its success with an event called 'Clean Rivers Commemoration' (Hon, 1990).

Table 6.3 Water quality parameters

Rivers	1978 (average values)			1985 (average values)		
	BOD	AMM-N	DO	BOD	AMM-N	DO
Singapore	11	1.7	NA	4	0.8	3.9
Bukit Timah/Rochor	58	13.0	NA	7	1.2	3.7
Kallang	76	26.5	NA	4	1.2	2.6
Whampoa	9	8.8	NA	3	0.5	3.4
Pelton	33	20.2	NA	3	0.8	3.2
Geylang	21	13.7	NA	7	1.1	3.2

Source: Chiang (1986).

Notes: DO = Dissolved Oxygen measures oxygen content in water, BOD = Biochemical Oxygen demand at 20 degrees Celsius and five days; it reflects of the amount of organic matter in the water and AMM-N = Ammoniac Nitrogen accounts for nutrients in water.

In 1955, the cost of cleaning the Singapore River was estimated at around $15 million, an amount that was revised upwards to $23 million in 1960 and to $30 million in an assessment carried out four years later in 1964. Considering that the final amount is calculated to have been anywhere between $200 and $300 million depending on the source, it can be observed how Singapore ended up paying many times the original estimate due to delayed efforts in carrying out this policy.[1] This is an important lesson for any government trying to improve the quality of their watercourses.

Overall, in terms of investment, Chou (1998) estimates the total costs incurred in carrying out the cleaning programme at $200 million. He also cites additional expenditures covered by the PSA, HDB and other government agencies as discussed earlier, of $21 million to form beaches in the Kallang Basin, $13 million in removing mud and other structures. Leitmann (2000) also puts the cleaning cost at $200 million, excluding the costs for public housing units provided, food centres, industrial workshops and sewerage. However, in a more recent study, Tan *et al.* (2009) set the clean-up costs at nearly $300 million, excluding resettlement compensation. It is not clear whether this figure included costs incurred directly and indirectly in manpower, time, education programmes in schools, public campaigns, etc. Whichever amount is used, it nevertheless reflects a sizeable investment in terms of financial, capital, human and opportunity costs. Nevertheless, it is one that continues to yield positive and oftentimes hard to fully quantify improvements to the general population's quality of life and the environment at large.

When comparing the costs of the river cleaning-up programme with the accrued benefits, one concludes that it was undoubtedly a worthwhile initiative. This scheme had numerous direct and indirect benefits, since it unleashed many development-related activities that improved the quality of life of the population, transformed Singapore's domestic landscape and international profile and enhanced its image as a model in urban planning and development. The value of land and demand for it along the waterways and catchments increased manifold and huge investments were made to attract tourism, recreation and related-business activities. For example, the areas of Boat and Clark Quays were soon renovated and turned into hub entertainment areas that combine modern and traditional structures. Similarly, Kallang Basin's new sandy beaches and parks transformed it into a location for water sports and other recreational activities, waterfronts, promenades and other commercial pursuits.

In addition to these area-specific achievements, once clean, the rivers have supported other long-term development plans. As part of a comprehensive and wide encompassing urban and national development scheme, buildings of architectural heritage along the banks of the Singapore River were restored and rehabilitated, helping the public landscape retain a sense of national identity. The entire programme left a legacy for future generations and gave the present one a refreshing sense of achievement. Moreover, the economic development of the land along the banks of the Singapore River, or the construction of a mass rapid transit tunnel under it, would have been impossible if the waterway and its surroundings had remained severely polluted.

The most important lesson, however, leads back to decisive, assertive and coordinated decision-making process. The exemplary political will of Singapore's leadership envisioned and encouraged a sustained process of social and economic progress and transformation through which the quality of life of the population was improved, successful urban development was achieved, the environment was protected, and the city-state placed itself on the right track towards sustainability.

Figures 6.1 and 6.2 show the Singapore River in 1971. Figure 6.3 shows the Singapore River in 2006. Figure 6.4 shows the Kallang Basin.

Figure 6.1 Singapore River, 1971 (photographers: John G. and Ma. Teresa Chamberlain).

Figure 6.2 Singapore River, 1971 (photographers: John G. and Ma. Teresa Chamberlain).

Figure 6.3 Singapore River, 2006 (credits to Public Utilities Board of Singapore)

Figure 6.4 Kallang Basin, 2010 (credits to Public Utilities Board of Singapore)

Further thoughts

When pondering why Singapore succeeded in planning, implementing and following up on its urban redevelopment programmes while other countries, even if highly industrialized, have not been able to do so, one would be right to conclude that such initiatives cannot be put in place in isolation. Singapore's experience shows that such schemes and efforts were attained because of the long-term scope in which they were envisioned, the comprehensive way in which they were planned, the effective institutional coordination behind them and the strong political will exerted by the country's leadership.

In the case of Singapore's urban development planning, this concurrently considered all relevant elements in a coordinated manner, for instance land acquisition, relocation and resettlement of affected families, provision of public housing, transportation improvements, etc., to mention only some. What is more, developing and strengthening the country's legal and institutional frameworks and technical expertise proved essential to the implementation of the different plans and programmes as was the cooperation and support of the general public. Planning was matched by implementation not because it was an easy task, but because the necessary frameworks were developed, adapted and pushed through by the city-state's leaders as deemed fit.

At the end of this process, dilapidated structures, temporary huts and hawker stalls gave way to good quality housing, commercial and industrial facilities, improved services, and recreational and leisure hubs with manicured pathways and lanes integrating old and more recent buildings, giving a new identity and character to the Singapore River. River banks, previously cluttered with boat-yards, backyard trades and squatter premises, have been transformed into beautiful riverside walkways and landscape parks, giving a new face to the city-state. The cleaning of the Singapore River and the Kallang Basin and the overall economic, social and environmental impacts this had represent only a small part of the overall vision to develop Singapore as a sustainable city. A vision many countries would do well to emulate.

As a visionary Prime Minister, Lee Kuan Yew, realized in the late 1960s, on a long-term basis it is much more expensive for a society to live in a polluted environment compared to a clean one. Nearly half a century later, most of the political leaders around the world are yet to recognize this fundamental fact.

7 Views of the media on the Singapore–Malaysia water relationship

Introduction

Singapore and Malaysia have established an intricate and closely interdependent relationship to which geography, historical heritage, economy and culture have contributed throughout the years. Singapore was a part of Malaysia between 1963 and 1965. After gaining independence in 1965, Singapore's lack of natural resources to support its economic growth and social development, mainly in terms of water resources, made the leadership of the country aware of the importance of developing and implementing clear visions, long-term planning and forward-looking policies and strategies that would provide it with enough flexibility to achieve its increasingly ambitious development plans (Ghesquière, 2007; Yap et al., 2010).

Historically, an important source of water for Singapore has been imported water from Johor[1] Malaysia, which will last, at least, until 2061. Four water agreements have been signed to that purpose: in 1927 (no longer in force), 1961 (no longer in force), 1962 and 1990. The 1927, 1961 and 1962 agreements gave Singapore the right to draw water from Johor and allowed Johor to buy treated water from Singapore in return. The 1990 agreement gave Singapore the right to build and operate the Linggiu Dam on the Johor River, and allowed Johor to sell to Singapore additional treated water generated from the Linggiu Dam over and above the quantum of water provided to Singapore by the 1962 agreement (see Appendix A). These agreements illustrate a history of consistent cooperation between the two countries on the issue of water, even when their 'water relationship' has not been exempt from serious disagreements and differences in opinion at different times in history (see Kog, 2001; Long, 2001; Kog et al., 2002; Lee, 2003a, 2005, 2010; MICA, 2003; NEAC, 2003; Chang et al., 2005; Saw and Kesavapany, 2006; Sidhu, 2006; Dhillon, 2009; Shiraishi, 2009; Luan, 2010).

Along the years, an important player in shaping the water relationship between the two countries has been the media. The media has acted both as a reporter and as a vehicle of communication, both officially and unofficially, to its own public but also to the interested parties in the other country. At some point, the media has been described as provocative on both sides (MICA, 2003; Chang et al., 2005), contributing to 'heighten emotions' in both countries (Chang et al., 2005: 3).

The fundamental importance of the media in regard to the water relationship between the two countries lies in the fact that, except for very few primary documents publicly available, such as those released by the Government of Singapore in 2003 (MICA, 2003), public information on the Singapore–Malaysia water relationship has been available primarily through the media. A clear indication is that the overall studies on this topic rely, sometimes heavily, on information disseminated initially in media reports.

This chapter analyzes the views and roles of the print media in the Singapore–Malaysia water relationship, mainly at the time when bilateral water negotiations were extensively covered (1997–2004). It also presents an overview of the media industries in both countries.

The management of the media in Singapore and Malaysia by the respective governments is beyond the scope of this chapter, and has been analyzed extensively elsewhere (e.g. Ang, 2002, 2007; George, 2007; Kenyon, 2007; Kim, 2001). Therefore, it is not discussed herewith.

The media in Singapore and Malaysia

The media industries in both Singapore and Malaysia are highly regulated by the respective governments. Characterized by numerous brands, but owned by a few companies, media content may not differ significantly from one medium to another. Both media structures are also described to have pro-government tendencies, which might have implications on the industry's attitude and reporting. While, ideally, coverage on the water relationship in both countries should be impartial, objective and factual, in real terms, reporting reflects to a significant extent the views held in the respective countries, a reason why portrayal of the events related to water negotiations may not necessarily be the same.

Singapore media

There are effectively only two print media companies in Singapore: Singapore Press Holdings (SPH) and MediaCorp. The SPH is a predominantly print media company, while MediaCorp, though mainly a broadcasting company, has one newspaper in circulation. The SPH has a 40 per cent ownership of the MediaCorp Press, and MediaCorp Press owns a substantial stake in SPH. Even though both companies are private groups, their management is linked to the government, generally holding a government-favourable stance (Ang, 2002; Tan, 2010).

Media market ownership in Singapore is often described as monopolistic (Ang, 2007; Tan, 2010). Except for *Today*, which is owned by MediaCorp Press, all print media is owned by SPH. The SPH publishes 17 newspapers in four languages and has 77 per cent of the readership in Singapore above 15 years old. The newspapers are as follows: *The Straits Times, The Sunday Times, The Business Times, The Business Times Weekend, The New Paper* and *The New Paper on Sunday* (in English); *Lianhe Zaobao, Zaobao Sunday, Lianhe Wanbao, Shin Min Daily News, zbComma* and *Thumbs Up* (in Chinese); *My Paper* (English and Mandarin); *Berita Harian* and *Berita Minggu* (Malay); and *Tamil Murasu* (in Tamil). *The Straits Times* is considered to be the most influential English newspaper in Singapore. In August 2010, it was the newspaper with the largest circulation, with 365,800 copies distributed daily (SPH, 2010).

The broadcasting media is dominated by MediaCorp, owned by Temasek Holdings, the government's investment arm. Internet-related media in Singapore is less restricted than print media, but is subject to controversial licensing regulations (Ang, 2007).

Media regulation in Singapore is extensive. The Newspaper and Printing Presses Act (Government of Singapore, 1974) requires publishers to renew licenses yearly. Media companies are also required by law to be public entities with no single shareholder controlling 12 per cent or more of a newspaper company without first obtaining government approval. Furthermore, the government has the legal authority to approve ownership and transfer of management shares that hold higher voting power in these companies. Publication that is 'contrary to the public interest' is prohibited under the Undesirable Publications Act.[2]

Malaysian media

In Malaysia, the United Malays National Organisation (UMNO) not only is the leading political party but also owns most of the main newspapers. Print media in Malaysia is in English, Malay, Mandarin and Tamil. The major companies are the New Straits Times Press (NSTP) and the Utusan Melayu Press (UMP). The NSTP publishes English newspapers, such as: the *New Straits Times, New Sunday Times, The Business Times*; as well as *Berita Harian, Berita Minggu, Harian Metro* and *Metro Ahad*. The Malay-based press holding UMP circulates *Utusan Melayu, Utusan Malaysia, Mingguan Malaysia* and *Utusan Zaman*. Besides the UMNO, other political parties, such as the Malaysian Indian Congress (MIC) and the Malaysian Chinese Association (MCA), are closely linked, in terms of ownership, to the media (Kim, 2001). Similar to Singapore, although there are numerous newspapers available in Malaysia, these are controlled by a few companies that are connected to the ruling party coalition (Shriver, 2003). In general, mainstream print media is not critical of the government (Kenyon, 2007).

Radio and television are also government-owned and controlled. The Internet, though still subject to some form of control, is the least restricted type of communication channel. However, the government is known to employ a surveillance system that restricts its access and has banned several websites (Kim, 2001). Through the Communication and Multimedia Act, licensing for Internet providers is also required and provides for legal actions against content that is considered to be defamatory and false.

The Internal Security and the Printing Presses and Publications Acts give mandate to the government in controlling the media. The Internal Security Act restricts coverage of matters that are considered a threat to national interest and security (Kenyon, 2007) and the Printing Press and Publications Act stipulates both granting and withdrawal of media licenses. Under this bill, where media publishers have to renew licences yearly, the government has the discretion to withdraw licences without any obligation for explanation (Kim, 2001).

Singapore–Malaysia water relations as viewed by the media

Before presenting the views of the media, it is important to introduce the situation regarding water negotiations during the 1997–2004 period, as addressed by the Ministry of Information, Communications and the Arts of Singapore (MICA, 2003) (see also Appendix B for a chronology of developments):

> The story began in 1998. Crises, they say, bring people together; so it was that at the height of the Asian financial crisis, the two countries began negotiations on a 'framework of wider cooperation'. Malaysia wanted financial loans to support its currency. To enable it to carry its domestic ground when acceding to the request, Singapore suggested that Malaysia give its assurance for a long-term supply of water to the Republic. Malaysia eventually had no need for the loans, and so negotiations turned to other matters of mutual interest. In particular, Malaysia wanted joint development of more land parcels in Singapore in return for relocating its railway station away from the current site at Tanjong Pagar.
>
> Over the ensuing three years, more items were bundled into the negotiation package. Singapore added one request: resumption of its use of Malaysian airspace for military transit and training. Malaysia added three more: replacing the Causeway with a bridge, early withdrawal of the Central Provident Fund savings for West Malaysians working in Singapore and a higher price for the water it presently sells to Singapore. Officials met; leaders corresponded; Singapore's Prime Minister Goh Chok Tong and Senior Minister Lee Kuan Yew took pains to visit Malaysian leaders at the capital, Kuala Lumpur.
>
> (MICA, 2003: 3)

The water negotiations came to the forefront of the Singapore–Malaysia cooperation framework as early as 1995, and in connection with talks on the Malayan Railway land in Singapore and Malaysia's proposal for an electric train that would connect both countries (*The Business Times*, 6 June 1997). By 1996, Malaysia's willingness to supply water to Singapore, contingent upon its domestic needs, was publicly announced (*The Business Times*, 6 June 1997). This was reinforced in the two-day visit of Singapore's Prime Minister Goh to Malaysia in February 1998, where both countries released a joint communiqué reaffirming this (*New Straits Times*, 4 April 1998 and Yang, 1998). The details of the new agreement were set to be refined within a 60-day period after the official visit. However, this deadline was not met because both parties could not reach an agreement on the details (*The Business Times*, 30 June 1998).

In 1998, the water negotiations were linked to a proposed financial assistance package from Singapore to Malaysia in the context of the economic problems experienced by Malaysia during the 1997 Asian financial crisis. Malaysia subsequently conveyed that it no longer had a need for the financial assistance. Malaysia then proposed the adoption of a 'package deal' approach (*The Business Times*, 18 December 1998 and MICA, 2003). This approach linked the issue of water to negotiations over other outstanding bilateral issues including:

> the Malayan Railway land in Singapore, the relocation of the immigration checkpoint for the Malayan Railway, the use of Malaysian airspace by Singapore aircraft, the transfer of Malaysian shares no longer traded on CLOB International to the Kuala Lumpur Stock Exchange, and the early release of CPF (Central Provident Fund) savings of Malaysian workers.
>
> (*The Business Times*, 18 December 1998)

There was very little media coverage of the status of the negotiations at that moment. Around the same time, Poh Lian Holdings, a Singapore-listed construction company, conducted feasibility studies to source water from Indonesia (En-Lai *et al.*, 2000).

In September 2001, Singapore Senior Minister Lee Kuan Yew and Malaysian Prime Minister Mahathir reached an in-principle agreement to resolve a host of outstanding bilateral issues, including water (*The Business Times*, 5 September 2001; Ng and Pereira, 2001). New to this agreement was the replacement of the causeway bridge in favour of a railway tunnel that would connect both countries (Ng and Pereira, 2001). Official proposals and counter-proposals were exchanged between the two countries, followed by a new round of bilateral talks between the two foreign ministers in July and September 2002. Several factors, such as Malaysia's unwillingness to discuss the terms of a new water agreement until a few years before the expiry of the 1962 agreement, and a disagreement over Malaysia's right to review the price of water sold under existing agreements, stalled the process. Subsequently, Malaysia delinked water from the package deal in October 2002. The package deal approach was dropped in favour of an individual approach for talks between senior officials in October 2002, but no deal was reached. Malaysia only wanted to discuss the current price of water, while Singapore also wanted the issue of future water supply included on the agenda (*The Straits Times*, 21 November 2002; MICA, 2003). The impasse led to discussions that circulated in the press about seeking legal recourse to resolve whether Malaysia had a right for a price review (*The Straits Times*, 21 November 2002).

In January 2003, Singapore released official letters from the negotiations in an effort to set the record straight (MFAS, 2003), following alleged misrepresentations and negative publicity about the negotiations in the Malaysian media. Several of these letters were published in Singapore's *Straits Times* (*The Straits Times*, 26 January 2003a, 2003b, 28 January 2003) and also were made available on the website of the Ministry of Foreign Affairs (MFAS, 2003). This decision was heavily criticized by Malaysia (Said, 2003a), who released, in July 2003, a week-long series of advertisements on the dispute titled 'Water: The Singapore-Malaysia Dispute: The Facts' (NEAC, 2003; Said, 2003b).

Shortly after, the official negotiations stopped. Malaysia stated that it would still honour current agreements but that negotiations were over. In contrast, Singapore expressed intentions of letting the first of the water agreements expire in 2011.

A host of additional factors also affected the bilateral water relations. These included the quest of both countries to be independent from each other in terms of water: Singapore wanting water self-sufficiency and Johor's aims to be independent of treated water from Singapore; the possibility for Singapore to source water from Indonesia; the development of the water industry in Singapore supported by the production of NEWater and desalination plants; reclamation efforts in Singapore which were allegedly affecting Malaysia, and the dispute over Pulau Batu Putih/Pedra Branca.[3]

Coverage of water relations

When analyzing the role of the media in the Singapore–Malaysia water relationship, it is fundamental to consider the prevailing historical and political context between the two countries at specific times. Regarding bilateral issues, experience has shown that media coverage, in general, does not necessarily follow the philosophy of the media as a public sphere where 'citizens discuss and deliberate matters of common interest and public concern, and hold the state accountable' (George, 2007: 94). In bilateral issues, media tend to be rather nationalistic with homogeneous views which focus primarily on the interests of their own countries and reflect the views of the respective states. The case of the Singapore and Malaysia media was no different.

The coverage of the water negotiations across the two countries evolved over time. At first, coverage was dominated by news articles portraying positive images of cooperation during the bilateral negotiations, and frames were consistent across Singaporean and Malaysian news. Later on, however, coverage grew increasingly negative and framing of the issues soon differed. The local interest grew with the proliferation of opinions and editorial articles alongside the news articles. Bilateral negotiations soon became a domestic political issue, and the leaders used the media to clarify and explain the status of negotiations to their respective public.

It is important to mention, at this point, that mass media coverage of political issues is normally, and necessarily, selective. The media depends on frames to give coherence to relatively brief treatments of complex issues through selective views, choosing to highlight certain items and ignore others, frequently looking for consensus or disagreement on certain issues. Since the number and variety of issues that an audience can appreciate on specific subjects is normally limited, public debate is constrained on matters even when they are relevant, simply because citizens tend to focus on specific issues when constructing their opinion. In the case of the Singapore–Malaysia water relationship – as one would expect to be the case in any normal bilateral relations – the media could decide to transmit mainly the views of their own government with a clear objective to form a favourable public opinion.

Jürgen Habermas, the German sociologist and philosopher, argued that the function of the media has been transformed from facilitating rational discourse and debate in the public sphere, into shaping, constructing and limiting public discourse to the themes approved by media groups (Kellner, 2000). Nevertheless, one could equally argue that citizens often trust their elites, and, in this case, also their national mass media, on bilateral issues, because they perceive them to be credible sources of information. In addition, as discussed by Andina-Díaz (2007: 66), the influence of the media is 'neither as significant as it was first thought to be, nor as minimal as was subsequently assumed'. This is because citizenry does not necessarily follow the media blindly or accept passively all views as presented, as they have their own motivations and own biases which may or may not coincide with the views presented by the media. Consequently, as important as the media is to shape public opinion, and as consistent as the messages of the media can be with their respective establishments looking to shape public opinion, a fundamental player in the equation is the readership. Citizens are not necessarily mere spectators who allow the media to mould their opinion: they normally make use of their own views and expose themselves to material with which they normally agree, often interpreting the media content to reinforce their own views and perspectives.

For this analysis, the period covered (1997–2004) for the bilateral negotiations has been divided into three parts following the media's evolving role (Table 7.1). The media's traditional role as a reporter of events is clear in the early phase of the water negotiations that is covered by the first period. The second and third periods include the increased role of the media as a communication platform. Nonetheless, while the second period demonstrates the media's role for clarification of facts for interested groups other than officials, the third period is identified by the media playing the role of an unofficial medium for communication between the governments of Singapore and Malaysia.

Each of these periods is discussed in detail in the following sections. This includes media portrayal from different news sources as well as the consistency

Table 7.1 The role of the media in the Singapore–Malaysia water relationship, 1997–2004

Media role	Period	Media content		Media framing
		Bilateral news	*Composition of print media*	*Comparative media coverage*
As reporter	First 1997–1998	Mostly positive content	News articles	Consistent
As reporter and platform for communication	Second 1999–2001	Increasingly negative content	News articles, opinions, editorials, forums and letters	Consistent with slight framing of news
	Third 2002–2004	Mostly negative content	News articles, opinions, editorials, forums, letters and pamphlets	Water negotiations were framed differently

of both media content (consistency of news content regardless of media source) and media framing of the news (assessment of the portrayal of the media content between different sources where the same media content may be presented or highlighted in different ways defining and constructing a political issue or a public controversy). This analysis also includes classifying the data as positive, negative or neutral, and making a comparison between the different media sources.

The media as a reporter

During the early phase of the negotiations (1997–1998), the media played largely a reporting role wherein news articles gave an account of the proceedings of the water negotiations between the two countries. Only the offices of the Prime Minister and the Foreign Minister of each country were directly involved in the negotiation process. In addition, there was a consistency in terms of the content that the Singapore and Malaysian media covered. This period mainly featured positive news coverage on the possibility of further cooperation on bilateral water agreements. Highlighted in these years were the countries' signing points of agreement (*The Business Times*, 6 June 1997) with Singapore willing to buy water from Malaysia and the latter willing to supply water to Singapore beyond 2061 (*New Straits Times*, 4 April 1998 and Toh, 1998). By mid-1998, there were some negative news items on the stalled talks due to a failure to reach an agreement between the countries, and a change from linking the water deal to a financial assistance package to a host of other bilateral issues (*The Business Times*, 18 December 1998). In spite of this, news coverage was still consistent in terms of content between Singapore and Malaysian and international media sources.

During this first period, media framing did not differ significantly between the two countries. However, there was a different choice of words between the Singaporean and Malaysian media with the same content being portrayed in different lights. For example, Singapore news reported that the UMNO youth 'calls for suspension of talks' (Toh, 1998), while, according to the Malaysian news, the UMNO youth 'urges the Malaysian government to be firm' in dealing with Singapore (Haron, 1998). Another difference was when Singapore news reported that Malaysia had 'agreed' to continue supplying water to Singapore (Wong, 1998b), while Malaysian news stated that it would 'not cut off water supply to Singapore' (*New Straits Times*, 5 August 1998). Regarding the loan assistance, Singapore news reported that Malaysia 'sought Singapore's help' (Wong, 1998b), which it eventually chose not to take (*The Business Times*, 18 December 1998), while Malaysian news reported that it had told Singapore that it 'does not need the funds offered by Singapore' (Abdullah, 1998b). Thus, even though media content in the two countries was fairly similar, the framing of the issues did differ across the causeway.

During these years, although there was consensus among the leaders to cooperate, there was no policy certainty, because no agreement could be reached. It is, thus, not surprising that in the first period, the media in both countries mainly played the role of informative reporter.

The media as a medium for communication

The media's role changed as it increasingly became a medium of communication between the two countries. During the second period (1999–2001), the media's role as a medium of communication was mainly for interested groups in the negotiation process other than official representatives. Finally, throughout the third period (2002–2004), the media's role further expanded when it served as an unofficial medium of communication between the two governments on the water negotiations. These aspects are discussed below.

The media as a medium of communication for stakeholders other than official representatives

During the second period, 1999–2001, the media played a second role. From a media coverage that was dominated by news articles, this period saw an increase in opinion and editorial articles that became gradually more negative in both countries.

There was still a general consistency in terms of media content in Singapore and Malaysia. The nature of the media coverage extended beyond just reporting, becoming a medium for communication and clarification. The media played a role in addressing concerns raised by both sides. For instance, Malaysian opinion articles portrayed the unfairness of the situation and how Singapore was benefiting from the existing water deal (Chin, 1999). Singapore opinion articles in turn portrayed how Singapore was not profiting from the deal while Malaysia was (*The Straits Times*, 3 March 1999; Tan, 1999). In addition, Johor's own interest in meeting its own water needs before Singapore's were also raised in the Malaysian Press (Venudran, 1999; *New Straits Times*, 8 June 1999). This was then addressed by the Singapore media, reiterating that the country's water demand was contingent upon Malaysia satisfying its own needs first (*The Business Times*, 11 June 1999). Thus, the media served as a platform for communication and clarification for the population of both countries.

Around this period, media coverage also voiced concerns on the over-dependence of Singapore and Malaysia on each other (*The Straits Times*, 2 December 2000; Toh, 2001). This might have contributed to other initiatives at this stage, where Singapore floated the idea of starting a partnership with Indonesia, led by the private sector, as an alternative source of water supply (Kagda, 2000, 2001; En-Lai, 2000; En-Lai *et al.*, 2000), and of investing in more desalination plants (Kuar, 2001a, 2001b). It was also reported that Malaysia had decided to build a water treatment plant in Johor to reduce reliance on receiving treated water from Singapore (Toh, 2000).

Given these events, media content was fairly consistent across the two countries and, even though it served to clarify and answer concerns raised by both sides, it covered roughly the same topics. Nevertheless, by then, the number of negative articles started to increase and the prolonged negotiations, which did not result in any agreement, may have facilitated this increased role of the media as an avenue for clarification. To some extent, this was to be expected given the increasing interest of the public in both countries on the status of the water negotiations. Compared to the first period, when only the official parties relayed information to the press, this interval saw various groups, such as Malaysia's UMNO youth group and Singapore's Workers Party, expressing their thoughts and views on the water negotiations (Toh, 1998; Haron, 1998; *The Straits Times*, 7 September 2001).

The exchanges between the Singaporean and Malaysian media, which were absent in the coverage of the international media, show the beginning of the framing of the water negotiations: Malaysia started portraying how the past water deals were in favour of Singapore (*New Straits Times*, 8 June 1999) and how Singapore was profiting from them (*New Straits Times*, 10 January 2001; Maharis, 2001), while Singapore started to portray how both countries had benefited from the agreements (Tan, 1999). Thus, within this role as a platform for communication and clarification, the media became a conduit to address the other country's concerns.

Though this period still projected an overall consensus among the leaders to cooperate, the degree of agreement began to change. For example, while there was a general consensus among the leaders, media coverage started to depict diverging and negative opinions. The prolonged negotiations were also an indication of heightened policy uncertainty. The media gradually became more focused on clarifying the respective government's views on various issues related to the negotiations, compared to its earlier role as a one-way (media to the public) reporter.

The media as an unofficial medium of communications

In contrast to the 1997–1998 and 1999–2001 periods when communication and clarification were the primary roles of the media, during the third period (2002 to 2004), the media's role changed even further. This is because, at that time, it started to serve as an unofficial medium for communication between the governments of Singapore and Malaysia. Furthermore, a distinct framing of the news was also more evident, where negative views were expressed on the Singapore–Malaysia bilateral ties. There was now a significant increase in the editorials, opinions and letters on the water negotiations.

The analysis below illustrates how the media played an unofficial role as a medium for communication between the two governments, and how the Singaporean and Malaysian media framed the news.

Unofficial medium for communication between governments

In one example, Foreign Minister Syed Hamid of Malaysia released news to the media about seeking legal recourse and cancelling negotiations with Singapore as early as October 2002 (Toh, 2002; Ahmad, 2002d; Abdullah, 2002a; Murugiah, 2002b). However, no official notice was forwarded to Singapore, with the Foreign Minister indicating that the newspapers knew where Malaysia stood in seeking legal recourse (Ahmad, 2002e). Another instance was the agenda of the negotiations on whether it would include both current and future water abstraction by Singapore or only current water requirements. Malaysia stated that it would only discuss the price of current water (Pereira, 2002c), while Singapore stated that both current and future water must be on the agenda (*The Straits Times*, 21 November 2002). Following the media reports, Singapore sought official clarification from Malaysia's Foreign Ministry to confirm the status of future water talks (*The Straits Times*, 1 December 2002). These examples illustrate how Singapore and Malaysia utilized the media as a platform for unofficial communication with each other.

In addition, the media played a role in delivering subtle signals and messages between the two governments. These signals captured the sentiments of both sides, which might or might not have come out during the official negotiations. For instance, when the Singapore media featured consecutive articles on NEWater's safety and quality (Nathan, 2002; *The Business Times*, 26 July 2002; Ming, 2002a, 2002b), Deputy Prime Minister Lee talked about the possibility of NEWater replacing the water imported from Johor (Teo, 2002), and Foreign Minister Jayakumar announced that water would no longer be a strategic vulnerability for Singapore (Latif, 2002). On the other side, Malaysian media also delivered subtle signals to Singapore when Foreign Minister Syed Hamid said that NEWater would not affect Malaysia's stand in the negotiations (*The Star*, 18 August 2002), and Prime Minister Mahathir informed the media that Singapore was free to stop buying water from Malaysia (Abdullah and Megan, 2002). Indirect signals, such as Johor officials negating the relevance of NEWater to Singapore's water security (*The Straits Times*, 11 August 2002), and a Malaysian MP's suggestion of selling sewage water to Singapore instead (*The Straits Times*, 3 October 2002), signalled Malaysia's position in the negotiations amid changing water dynamics.

Therefore, the media played an unofficial platform not only for communication on bilateral negotiation topics, such as the agenda and arbitration, but also for sending subtle signals to both countries. With the growing interest in the water negotiations, the media also served as a platform for both parties to explain and educate their domestic constituencies about the progress of the negotiations. The speech of Singapore's Foreign Minister Jayakumar to the Parliament detailing the water talks was published in the newspapers (*The Straits Times*, 26 January 2003c). Simultaneously, several of the letters from which the Foreign Minister quoted extensively were released to *The Straits Times* (*The Straits Times*, 26 January 2003a, 2003b, 28 January 2003). These letters were also archived and posted on the Ministry of Foreign Affairs website (MFAS, 2003). Malaysia published a week-long series of advertisements on the issue, stating its position in the water talks (Said, 2003b). This information was also compiled and made available on Malaysian government websites (NEAC, 2003). Furthermore, Malaysia made these copies available for sale to the public (Abdulla, 2003a). It is in these publications that the Singaporean and Malaysian media's framing of the water negotiations became more evident.

Media framing of the Singapore–Malaysia water relations

Both Singapore and Malaysia portrayed different views of the water negotiations that became most evident in the third period. These views were valid, but incomplete when viewed and read separately. An informed public needs to understand both sides of the situation. More often than not, however, the media seem to have reported what the two governments wanted their respective public to know about their respective policies.

Not surprisingly, both the Singaporean and the Malaysian governments portrayed themselves as the more reasonable partner in the bilateral talks. Each party claimed that the other was publishing inaccurate information. Malaysia claimed that Singapore published false information on the water issue (*Berita Harian*, 26 July 2003), and that its reports were misleading and did not accurately reflect the developments (*Sin Chew Daily*, 29 January 2003). Meanwhile, the Singaporean media reported how Malaysia ignored crucial facts (Lee, 2003) and published false answers in its media blitz (*Berita Harian*, 28 July 2003).

The divergent framing of the critical issues from Singapore and Malaysia is discussed in the next section.

The Singaporean media on water relations: Singapore is consistent and reasonable

The Singaporean media portrayed Singapore as a consistent and reasonable negotiating partner in the water negotiations. It repeatedly framed Singapore as very accommodating, giving in to Malaysia's changing requests (Ming, 2002c; Latif, 2002). It highlighted how Malaysia had been inconsistent and unreasonable in the negotiations: changing their nature, changing the agenda and changing the price of water.

Changing the nature of negotiations. The extension of the water contracts was first tied to a financial arrangement (Kassim, 1998a; Teo, 1998; Wong, 1998b). It was then that Malaysia requested to drop the financial issues and instead adopted a package-deal approach to include a host of outstanding bilateral issues between the two countries (Abdullah, 1998b; *The Straits Times*, 24 July 2002). This approach linked the water issues to the early withdrawal of the Central Provident Fund for Malaysians, the relocation of the Malaysian Immigration checkpoint in Singapore and the use of Malaysian airspace by Singapore (*The Business Times*, 18 December 1998). It was also Malaysia that added new issues into the agenda, such as the proposal for a new bridge to replace the Causeway (*The Straits Times*, 26 January 2003c; How, 2002). In the end, it was also Malaysia that unilaterally dropped the package approach (Hoong, 2002; Toh, 2002c). Even then, Singapore was still willing to continue negotiations with Malaysia (*The Straits Times*, 24 July 2002 and 26 January 2003c).

Changing the agenda of water negotiations. Focusing specifically on water, Malaysia often changed the rules. The original talks were on the extension of the water agreements after the 1961 and 1962 contracts had expired (*The Business*

Times, 30 June 1998; Wong, 1998b; *New Straits Times*, 4 April 1998; Abdullah, 1998a). Malaysia was the party that wanted to include the prevailing agreements on the table through a price revision for water, and for the price review to be retroactive (Hoong, 2002; *The Straits Times*, 24 July 2002; Toh, 2002c). Later on, it was also Malaysia that wanted to discuss only the price of current water and not of future water (Low and Toh, 2002; Toh, 2002e; Pereira and Lim, 12 October 2002).

Changing the water price. Malaysia constantly changed its offer as to the price Singapore should pay for importing water (*The Business Times*, 1 February 2002; Tan, 2002; Toh, 2002d; Ahmad, 2002a). From an agreement of 45 sen per 1,000 gallons in 2000, Malaysia changed its asking price to 60 sen per 1,000 gallons in 2001 and then to RM3 per 1,000 gallons (*The Straits Times*, 26 January 2003c). Together with the shift from a package approach to dealing with the water issue separately, the changing water price added to the protracted negotiations on water. It is important to mention that Singapore is subsidising the cost of treated water to Johor. Singapore is selling treated water at the rate of RM0.5 per 1,000 gallons, although it is costing Singapore RM2.4 per 1,000 gallons.[4]

The Singapore media reported how Singapore had been reasonable and accommodating to Malaysia's changing requests (Ming, 2002c; Latif, 2002). However, the alleged misrepresentations and negative publicity about the negotiations in the Malaysian media prompted Singapore to clarify the issues by publishing official correspondence between the parties concerned (*The Straits Times*, 26 January 2003c). The speech of Foreign Minister Jayakumar and the additional information released through the letters further indicate how the media had framed the situation.

The Malaysian media on water relations: Malaysia wants a fair price

The Malaysian media portrayed Malaysia as the most reasonable partner who only wanted a fair water agreement with Singapore (*New Straits Times*, 21 July 2003). Likewise, the publication of its booklet *Water: The Singapore–Malaysia Dispute: The Facts* (NEAC, 2003) also solidified this framing. This booklet highlighted how Malaysia had been a cooperative partner, willing to supply water to Singapore as long as a fair agreement was made.

Willingness to supply water to Singapore. Malaysia repeatedly expressed its sincerity in supplying water to Singapore even after the 1961 and 1962 agreements had expired (*New Straits Times*, 4 April and 5 August 1998; Osman, 2002a; Hong and Abdullah, 2003). The governments of Pahang and Johor also expressed the same view (Aziz and Haron, 1998; *The Straits Times*, 21 January and 1 October 1999).

Desire for a fair agreement. Malaysia was just requesting a fair and reasonable arrangement with Singapore and asking for a fair price (*New Straits Times*, 19 and 21 July 2003; Said, 2003c; Sooi, 2002a). Instead, Singapore had been profiting from Malaysia due to the low cost of water supplied by Malaysia and the high price Singapore was charging for treated water subsequently supplied to Malaysia (*New Straits Times*, 13 March 2000, 10 January 2001; Said, 2002). To make progress, Singapore should accept Malaysia's right for a price review (Osman, 2002c; Said and Cruez, 2003).

Reasonable Malaysia, unbending Singapore. It was Singapore that was being unreasonable and refusing to accommodate Malaysian requirements which were fair and just (*New Straits Times*, 13 March 2000). Singapore repeatedly turned water into the pivotal issue in negotiations that forced Malaysia to delink water from the package approach (*New Straits Times*, 23 and 26 January 2002; Said, 2002; Seong, 2002).

The Malaysian media reported how Malaysia was very reasonable towards Singapore, only asking for a fair price for its water. To counteract Malaysian aims, Singapore published official correspondence between the leaders of the two nations. This forced Malaysia to react and inform the Malaysian population on its views of the situation through an information campaign.

The media, its views and the water relations

When discussing the role of the media in the Singapore–Malaysia water relationship, it is essential to understand the historical and political context between the two countries at specific times.

Historically, water has always been considered a very important part of the Singapore–Malaysia bilateral agenda and, as such, water agreements have been signed in different years between 1927 and 1990. Later on, water became part of the so-called 'package deals', where discussions included several other issues. Along the years, one would have expected that the historical and political situations prevailing between the two countries had influenced the views and opinions as well as the 'sentiments' and even the 'tone', of the media when covering the news, even if these were supposedly solely about water.

Overall, the main role of the media has been to publicize and inform the public on the water negotiations, mainly from the viewpoints of the respective national interests, which is to be expected on bilateral issues. This publicity has been effective to the point that other groups, both private and public, that were not party to the official negotiation process, expressed their own thoughts and opinions on what their governments should do. For example, there was an increase in the number of interest groups, such as the UMNO youth group, the PAP (People's Action Party) (Yap *et al.*, 2010) youth group, opposition parties, NGOs, the Malaysian military and research institutions, that expressed their opinions on the issue. The increased media coverage and the larger number of interested stakeholders even prompted the Singaporean and Malaysian Foreign Ministries to explain the details of the protracted talks to their citizens. The complexity created by the emergence of so many new players even impelled the Johor Menteri Besar to ask the Malaysian public to stop interfering in this matter (*Sin Chew Daily*, 20 January 2003).

By and large, the local interest in the two countries grew with the proliferation of opinion and editorial articles. When bilateral negotiations became a domestic political issue, the leaders used the media to clarify and explain the status of nego-tiations to their respective public, trying to make their views clear for their own citizens, but also sending 'unofficial' messages to interested parties in the other country. Clearly, with the growing interest in the water negotiations, the media also served as a platform for the countries not only to inform their citizens but also to educate them on the progress of the negotiations.

As mentioned earlier, mass media coverage of political issues is normally, and necessarily, selective. The media normally depends on frames to give coherence to relatively brief treatments of complex issues through certain views, in which they select to highlight specific items and ignore others, often looking for consen-sus or disagreement on certain matters. Nevertheless, as important as the media is to shape public opinion, and as consistent as the messages of the media can be with their respective establishments, citizens normally make use of their own views, listening to some information and disregarding some other, depending on whether they agree or not with the views presented. It so happens that citizens often interpret media content to reinforce their own views.

The view on the role of the media can be more comprehensive if considered within its multifaceted relationship with the state and the public, with the three actors being equally important in the equation. In the water relationship between Singapore and Malaysia, the media has been a dynamic actor – a role which has evolved with time during the course of the bilateral negotiations – but also has taken on the attitude of the citizens in both countries. While the media of each country has portrayed the water negotiations in different lights, playing impor-tant roles in terms of reporting on policies and politics, the readers have also played an important role mainly in terms of supporting their own countries, often willingly accepting the viewpoint of the establishment, considering it to be acting according to their national interests. The role of the media would be partial only if considered in isolation, without acknowledging the role of the readers, since, frequently, both of them voice the ideas, ideals and concerns of the other.

Further thoughts

Relationships between Singapore and Malaysia have 'undergone a sea change' during the last decade or so (Chang *et al.*, 2005: 1), with bilateral relations improving significantly with the change of leadership in Malaysia in 2003. Stronger bilateral ties have been characterized by greater contact and coopera-tion between the leaders, officials and businesses. Synergies are developing and include, but are not limited to, economic cooperation, trade and investment; increasing private sector participation in strategic investments, corporate pur-chases and joint business ventures; cooperation on security matters; movement of technical experts across borders; promotion of tourism and sport-related activ-ities; exchange of students; and improved relationships between civil society groups in both countries (Saw and Kesavapany, 2006; Sidhu, 2006). With the objective to promote better understanding and bilateral ties among the citizens of both countries, circulation of newspapers on 'both sides of the Causeway' has been re-initiated after a 30-year ban of each other's newspapers (Saw and Kesavapany, 2006: 17).

Bilateral negotiations also resumed between 2004 and 2006 with the aim to solve several outstanding issues. In 2004, it was agreed, by Singapore's former Prime Minister Goh and then Malaysian Prime Minister Abdullah Badawi, that future discussions between the two countries should be based on the consideration of mutual benefits on any proposal that would be discussed. It was also agreed that the issues that had not been resolved should not delay cooperation in other areas (Saw and Kesavapany, 2006). Although this round of negotiations did not lead to a deal, a new phase in relations where both countries wanted to resolve bilateral differences had begun. The more positive political environment has already resulted in the amicable solution of several outstanding issues. For example, the two sides settled a land reclamation dispute in 2005. Among the terms of the settlement, Malaysia withdrew its case against Singapore from the International Tribunal of the Law of the Seas, while Singapore agreed to make adjustments to its reclamation works, and compensated Malaysian fishermen for losses due to the works. Both countries signed an agreement, on 26 April 2005, that the Johor Straits form a 'shared water body' (Sidhu, 2006: 88). Malaysia has expressed many times its intention to replace the causeway that links the two countries (and under which the water pipes run from Malaysia to Singapore). At some point, Malaysia talked about a 'scenic bridge' that would replace the causeway. The dispute on the sovereignty of Pedra Branca was resolved by the International Court of Justice in 2008. Regarding the resolution of the dispute over interpretation of the 1990 Points of Agreement (POA) on Malayan Railway land, both leaders discussed issues arising from the POA at the Singapore–Malaysia Leaders' Retreat in 2010 and came to an agreement to move the issues forward by supplementing the POA with new terms and conditions (MFA communication).

In terms of the media, it should be noted that Singapore and Malaysia resumed talks on outstanding bilateral issues in 2005, and decided not to divulge the details of the negotiations through the media (MFAS, 2005). Both countries have agreed that details of the negotiations on water should not be discussed with the media. It has been recognized that it would not be helpful to publicize the details of the negotiations to avoid heightening expectations, as happened on earlier occasions, as well as to avoid a media frenzy on whatever issues had been discussed, as had been the case in the past (and would be the case once more in the media reports on the 'scenic bridge' case presented in Saw and Kesavapany, 2006: 6–7).

Former Singapore Prime Minister Goh is quoted as saying that 'due to the sensitive nature of issues both sides have agreed to keep discussions in private instead of negotiating through the press—as it has been the case in the past' (Agence France-Presse, 17 October 2004, in Sidhu, 2006: 87). The low-key and private nature of the discussions that have been held are considered as clear signals of the willingness of both countries to solve bilateral problems between them, as well as to avoid the media capitalizing on these issues, as it did in the past (Sidhu, 2006). A private setting or 'quiet diplomacy' has clearly been accepted by both countries as the best way to achieve progress (Lee, 2010).

The ties between both the countries are deep-rooted and are based not solely on water but on a multiplicity of other factors. As noted by Tan Gee Paw, current Chairman of the Public Utilities Board of Singapore, 'There is much that both countries can gain by working together. Our common interests far exceed our bilateral differences' (BBC World Debate on Water, Singapore, 30 June 2010).

Appendix A

Singapore–Malaysia water agreements

The water relationship between Singapore and Malaysia dates as far back as 1927, decades before both countries gained independence. Even though the water relations have not always been perfect, the four water agreements signed in 1927, 1961, 1962 and 1990 illustrate a history of consistent cooperation between the two countries on the issue of water. These agreements have secured a supply of water for Singapore, allowed Malaysia to buy treated water from Singapore, and brought about development in Malaysian water infrastructure through Singaporean investments in Johor. Table 7A.1 summarizes the water agreements between Malaysia and Singapore.

Table 7A.1 Water agreements between Singapore and Malaysia

Water agreements	*Parties to the agreement*	*Water agreement details*
1927	The Sultan of Johore and the Municipal Commissioners of the Town of Singapore	The agreement allows Singapore to rent 2,100 acres of land in Gunong Pulai at 30 cents per acre per year, and 'take, impound and use all the water which from time to time may be or be brought or stored in upon or under the said land' at no cost to the municipality. Singapore also had the right to lay and maintain the necessary waterworks to transfer the water. The Government of Johore could request the supply of 800,000 gallons of water per day, if necessary, at 25 cents per 1,000 gallons.
1961	The Johore State Government and the City Council of the State of Singapore	Under this agreement, the Government of Johore reserved the lands, hereditaments and premises situated at Gunong Pulai, Sungei Tebrau and Sungei Scudai in the State of Johore for the use by the City Council. The City Council shall pay to the Government an annual rent of $5 per acre. The Government of Johore shall not for a period of 50 years alienate or do any act of deed affecting the said land or any part thereof during such term. The Government of Johore agreed to give the City Council the full and exclusive right and liberty to enter upon and occupy the land, and take, impound and use all water from the Tebrau and Scudai rivers, as well as to construct the necessary water works, reservoirs, dams, pipelines, aqueducts, etc. The City Council would supply to the Johore Government, if

Table 7A.1 (Continued)

Water agreements	Parties to the agreement	Water agreement details
		and when requested by the Government, a daily mount of water not exceeding at any time 12 per cent of the total quantity of water supplied to Singapore over the Causeway, and in no case less than four million gallons. The quality of the water would have to be of accepted standard and fit for human consumption. The City Council would pay to the Government three cents for every 1,000 gallons of water drawn from the State of Johore and delivered to Singapore, and the Government of Johore would pay to the City Council 50 cents for every 1000 gallons of pure water. When the City Council had to provide the Johore Government with raw water, it would pay 25 cents for every 1000 gallons of the water supplied. These clauses are subject to review after the expiry of 25 years time. Prices can be revised in the light of any change in the purchasing power of money, cost of labour and power and material for the purpose of supplying the water. In the event of any dispute or differences arising under the provisions of this clause the same shall be referred to arbitration as provided in the agreement.
1962	The Johore State Government and City Council of Singapore	The Government of Johore agreed to demise unto the City Council all and singular specific lands in the State of Johore for a period of 99 years. The Government granted the City Council 'the full and exclusive right and liberty to draw off, take, impound and use the water from the Johore River up to a maximum of 250 million gallons per day'. The City Council would supply to the Government a daily amount of water drawn off from the Johore River not exceeding at any time 2 per cent of the total quantity of water supplied to Singapore, the quality of which would always be of acceptable standard and fit for human consumption. The City Council shall pay to the Government three cents for every 1,000 gallons of water drawn from the Johore River and delivered to Singapore. The Government would pay to the City Council 50 cents for every 1,000 gallons of pure water supplied. Should it be necessary for the City Council to supply raw water to the Government, this would pay to the City Council ten cents for every 1,000 gallons of the raw water. The price can be revised after the expiry of the agreement in 25 years' time, and in line with the rise or fall in the purchasing power of money, cost of labour, and power and materials for the purpose of supplying the water. In the event of any dispute or differences arising under the provisions of this clause the same shall be referred to arbitration as provided by the agreement.
1990	The Government of the State of	The Johor Government agreed to sell treated water generated from the Linggiu Dam to PUB in excess of

Johor* and the
Public Utilities
Board of the
Republic of
Singapore

the 250 million gallons per day of water under the 1962 Johor River Water Agreement considering that PUB agreed 'to build at its own cost and expense the Linggiu Dam and other ancillary permanent works in connection therewith and thereafter to run, operate and maintain at its own cost and expense the dam, reservoir and ancillary permanent works'. The Agreement shall expire upon the expiry of the 1962 Johore River Water Agreement. However, it can be extended beyond the original terms should the parties agreed to it. It was agreed that PUB shall purchase treated water 'at the price of either the weighted average of Johor's water tariffs plus a premium which is fifty per cent of the surplus from the sale of this additional water by PUB to its consumers after deducting Johor's water price and PUB's cost of distribution and administration of this additional water, or 115 per cent of the weighted average of Johor's water tariffs, whichever is higher'. The quality of the treated water supplied to PUB under this Agreement shall conform with the prevailing World Health Organization's guidelines for drinking water. In terms of land, the Johor Government agreed that the State land to be used for the catchment area and the reservoir of approximately 21,600 hectares shall be leased for the remaining period of the 1962 Johore River Water Agreement. The premium for the land shall be calculated at the rate of M$18,000 per hectare and an annual rent at the rate of M$30 for every 1,000 square feet of the said land. The annual rent will be subject to any revision imposed by the State Authority under the provisions of the National Land Code of Malaysia (Act No. 56/65). PUB agrees to pay M$320 million as compensation for the permanent loss to the use of the land, the loss of revenue from logging activities in the form of premium, royalty and cess payment and the one-time up front payment for the leasing of the said land, inclusive of rentals for the remaining tenure of the 1962 Johore River Water Agreement. Any dispute or difference between the parties which cannot be resolved amicably by discussions between the parties shall be settled by arbitration in accordance with the Rules of the Regional Centre of Arbitration at Kuala Lumpur at that time.

Sources: see the several water agreements and the Constitutions of Singapore and Malaysia for further information: The Agreement as to Certain Water Rights in Johore between the Sultan of Johore and the Municipal Commissioners of the Town of Singapore signed in Johore on 5 December 1927; The Johore River Water Agreement between the Johore State Government and City Council of Singapore signed in Johore on 29 September 1962; The Tebrau and Soudai Rivers Agreement between the Government of the State of Johore and the City Council of the State of Singapore signed on 1 September 1961; Agreement between the Government of the State of Johor and the Public Utilities Board of the Republic of Singapore, signed in Johore on 24 November 1990; Guarantee Agreement between the Government of Malaysia and the Government of the Republic of Singapore signed in Johore on 24 November 1990.

* Spelling as in the Agreement.

Appendix B

Table 7A.2 Chronology of key developments

17 December 1998	PM Goh agreed with PM Mahathir's proposal to resolve outstanding bilateral issues, including long-term supply of water to Singapore, together as a package.
March–May 1999	Officials from both sides met three times, but made little progress.
15 August 2000	At a four-eye meeting in Putrajaya, SM Lee and PM Mahathir reached an agreement on a list of items including the price of 45 sen per 1,000 gallons for current and future water. This was the first time the issue of current water was discussed as part of the package. Singapore also agreed to discuss Malaysia's proposal to build a new bridge to replace the Causeway as part of the package.
24 August 2000	SM Lee wrote to then Malaysian Finance Minister Tun Daim Zainuddin confirming the list of items which he and PM Mahathir had agreed to.
21 February 2001	PM Mahathir replied to SM Lee that 'Johore believes that a fair price would be 60 cents [sic] per mgd [sic] of raw water' and that this 'should be reviewed every five years' (he meant 60 sen per 1,000 gallons).
23 April 2001	SM Lee noted in his reply to PM Mahathir that there were two main variations from their oral understanding reached in August 2000. These were PM Mahathir's proposal of 60 sen for raw water and the mix of raw and treated water to be supplied.
4 September 2001	SM Lee met with PM Mahathir for a second time in Putrajaya. At a joint press conference, they announced that they had agreed on the basic skeleton of an agreement. SM Lee explained that Singapore had offered to pay 45 sen for current raw water in return for assured raw water supply from Malaysia beyond 2061.
8 September 2001	SM Lee wrote to PM Mahathir to follow-up on their 4 September 2001 discussion on Malaysia's proposal for a bridge to replace the Causeway.
21 September 2001	SM Lee wrote again to PM Mahathir concerning other issues in the package. Singapore affirmed its proposals to revise the price of current water from three sen to 45 sen per 1,000 gallons, in return for Malaysia agreeing to supply water, at 60 sen, beyond the expiry of the existing agreements, in 2011 and 2061. The 60 sen price would be reviewed every five years for inflation.
18 October 2001	PM Mahathir now said Johor wanted 60 sen for water sold to Singapore. He also suggested that Singapore compensate Malaysia with more land parcels, should the KTM rail service end in Johore Baru.
10 December 2001	SM Lee replied to PM Mahathir to clarify Singapore's proposal on the bilateral issues and to seek clarification on the additional railway lands

referred to by PM Mahathir. He expressed the hope that PM Mahathir would consider the long-term significance and value of retaining the railway link between Malaysia and Singapore. He requested PM Mahathir to set out Malaysia's position on the package of issues so as to establish a clear framework for officials to work on.

5 February 2002	Prompted by repeated Malaysia comments to the media that existing water agreements were unfair, Singapore conveyed a Third Party Note to register its deep concern over those remarks.
4 March 2002	PM Mahathir conveyed yet another new pricing proposal for water – this time a three-stage proposal. The asking price was now 60 sen for water from 2002 to 2007, RM3 from 2007 to 2011, and RM3 adjusted for inflation every year after 2011. As for the treated water Johor now buys from Singapore, Malaysia proposed that the price be raised simply from the current 50 sen to RM1, with no price review mechanisms.
11 March 2002	SM Lee wrote to PM Mahathir noting that the latest proposals had changed completely from those agreed upon early. He would therefore have to study the implications of Malaysia's new offers before responding.
11 April 2002	PM Goh wrote to PM Mahathir, conveying Singapore's response to the latests proposals. He said that for the sake of good long-term relations, Singapore would supplement the existing water agreements by producing its own NEWater. Since Malaysia did not accept Singapore's earlier offer of 45 sen for current water and 60 sen for future water, Singapore proposed to peg the price of future water to an agreed percentage of the cost of alternative source of water, which was NEWater. He also suggested that PM Mahathir's letter of 4 March 2002 and PM Goh's reply of 11 April 2002 form the basis for further discussions between the respective Foreign Ministers and officials.
1–2 July 2002	The two Foreign Ministers and their officials met on Putrajaya. Malaysia invoked the Hongkong (spelling in the printed text) model, in which Hongkong pays China RM8 per 1,000 gallons for its water. Singapore said it was willing to negotiate a price review provided this is done as part of a package even (t)hough it believes Malaysia's right to review expired in 1986 and 1987. It also pointed out that unlike Singapore, Hongkong does not have to bear the infrastructure and maintenance costs of drawing water.
2–3 September 2002	The Foreign Ministers met a second time, in Singapore. This time, Malaysia proposed a formula that resulted in a price of RM6.25 per 1,000 gallons for current raw water. Malaysia also proposed that discussions on future water take place only in 2059.
8 October 2002	PM Mahathir told PM Goh while both were in Putrajaya that Malaysia wanted to 'decouple the water issue' from the other items in the package. PM Goh responded that if the water issue was taken out of the package, then Singapore would have less leeway to make concessions on other issues.
10 October 2002	PM Goh received a letter from PM Mahathir dated 7 October 2002, in which Malaysia declared that it had decided to discontinue the package approach. Dr Mahathir did not mention that he had written this letter when he spoke to Mr Goh on 8 October 2002.
14 October 2002	PM Goh replied to PM Mahathir. He noted that since Malaysia wanted to discontinue the package approach, Singapore would have to deal with water and the other issues on their stand-alone merits and no longer as a package.

16–17 October 2002	Senior officials from both sides met in Johor Bahru to discuss the water issue. But Malaysia wanted only one aspect of water discussed – the price of current raw water. Singapore reiterated that Malaysia had lost its right to the review, but it would agree if Malaysia agreed also to discuss the supply of future water. Singapore also asked Malaysia to explain how it had arrived at the price of RM6.25 it was asking for. It said that going by the terms of the water agreements, any review would result in a price of not more than 12 sen in 2002. Malaysia could not provide a satisfactory explanation.

Source: MICA (2003, Annex C, pp. 82–83).

Note: The version available at http://app.mfa.gov.sg/data/2006/press/water/event.htm is somewhat different to the printed version in the wording of the events and also in the detail in which it describes the several events. The version available online also includes comments on two meetings on 14 and 25 March 2002, which are not included in the printed version and therefore are not included in this table.

Appendix C
Data summary for the role of media
in Singapore and Malaysia water relations

Table 7A.3 Media as a reporter of events in the first period, 1997–1998

	Singapore News	Malaysia News	International
Media content	1. Agreement has been reached between the two countries and the use of a wider cooperation framework may be underway.	1. Both countries have reached an understanding on the water issue but no agreement has been signed.	1. Malaysia will provide water to Singapore given conditions.
	2. UMNO youth calls for the suspension of dealings with Singapore and a review of water supply agreements.	2. UMNO youth urges Malaysian government to be firm in defending its position on issues related to bilateralties and to fix a more reasonable price for the water supply agreements.	2. Malaysian leader fires hot words and threaten to cut water supply.
	3. Singapore is still keen to buy water from Malaysia. Malaysia agrees to continue supplying water to Singapore beyond 2061.	3. Prime Minister Mahathir assures that Malaysia will not cut water supply to Singapore.	4. Prime Minister Mahathir criticizes Singapore with Lee Kuan Yew's book release.
	4. Malaysia has sought help from Singapore to raise funds to tackle the country's economic crisis. Malaysia then did not take the loan facility from Singapore and proposed to link the issue of water on all bilateral problems.	4. Malaysia has told Singapore that it does not need the RM15.2 billion offered by Singapore as funds-for-water agreement; and proposes all outstanding issues to be discussed as a package.	5. Countries have resolved to sink differences to resolve outstanding issues.
Media framing	1. Malaysia *seeking* help from Singapore.	1. Malaysia *does not need* the financial assistance package.	
	2. Singapore news stating the Malaysia *agrees* to continue supplying water.	2. Malaysia assures Singapore that it will *not cut off* water supply.	

Source: authors' compilation from print media data. Italics are the authors' for emphasis.

Table 7A.4 Media as a medium of communication in the second period, 1999–2001

	Singapore	Malaysia	International
Media content	1. Malaysia says that Singapore should buy water from them because it is cheaper than in Indonesia. Pahang is still keen to sell water to the republic. 2. Singapore's water demand is always contingent on Malaysian (thus including Johor's) water needs. 3. Singapore is not profiteering from the water deals. In fact, Malaysia benefits from it. 4. Malaysia has allocated investment for water treatment plant in Johor. Forum articles write that Malaysia can choose not to buy water from Singapore. Forum articles also write that Singapore must end its reliance on Malaysia and Indonesia and consider alternative sources of water. 5. Both countries have reached an in-principle agreement under a package deal approach. Singapore has conceded to Malaysia in this deal.	1. Malaysia will supply Singapore with treated water after 2061. 2. Johor should first safeguard its own interests. It must secure local water needs first. 3. Malaysia was short-changed in the past water deals because they favoured Singapore. The republic is profiteering from these deals. 4. Johor will soon stop buying treated water from Singapore. It will soon build its own water treatment plant. 5. Johor also benefits from the water agreements.	1. Malaysia and Singapore will discuss bilateral issues as a package. 2. Singapore refutes Malaysia's profiteering claim. 3. Singapore Minister ruled out the possibility of water cut-off from Malaysia and stressed the need to diversify water sources 4. Singapore and Malaysia make an effort to resolve deadlock issues and reach an in principle agreement that marks a psychological breakthrough.
Media framing	Water agreements were beneficial to both parties and Singapore is not profiteering from water deals.	Malaysia was short-changed and Singapore is now profiteering from water deals.	

Source: authors' compilation from print media data.

Table 7A.5 Media as an avenue for negotiations in the third period, 2002–2005

	Singapore	Malaysia	International
Media content	1. Malaysia has lost its right to review the price of water. The sanctity of the agreement and Singapore's sovereignty is at stake. 2. Singapore denies unfair deal in water. It subsidizes the treated water sold to Johor. 3. Malaysia wants a higher price for water. 4. Malaysia can't simply alter water supply deal with Singapore because it is bound by the terms of the pact. Malaysia will resort to arbitration to settle the water issue with Singapore. 5. Singapore is not against an upward revision of the current price but that there must be basis as to what the new price should be. 6. Malaysia has unilaterally removed the negotiations on the water agreements from the package of outstanding issues. Thus all Singapore concessions are off. 7. Singapore releases letters to set things clear. It also published *Water Talks: Only if it Could* to counter Malaysian claims.	1. Malaysia wishes to invoke its right to review the price of water. 2. Singapore has turned the water issue into a pivot for the package deal and is holding all issues hostage. 3. Singapore is free to stop buying water from Malaysia if it wishes to. 4. Singapore has consented to discuss the water issue alone. 5. Malaysia has decided to stop negotiations and seek legal recourse instead. 6. Malaysia questions Singapore's action in making public the correspondence between the leaders of both countries. 7. Malaysia releases advertisements in local newspapers on water relations. It also published a compiled version *Water: The Singapore–Malaysia Dispute: The Facts*	1. Malaysia and Singapore is unable to agree on water supply price. 2. Singapore denies Malaysian allegations of water agreements favouring the republic. 3. Malaysia and Singapore agree to take the water issue separately. 4. Singapore blames Malaysia for breakdown on water talks. 5. Malaysia defends public message on Singapore water issue and published *Water: The Singapore-Malaysia Dispute: The Facts*.
Media framing	Singapore is consistent and reasonable throughout the water negotiations. It is Malaysia that has been inconsistent, changing the nature of negotiations, the agenda of water negotiations and the water price.	Malaysia is a reasonable party wanting a fair water agreement with Singapore. Malaysia has been willing to supply water to Singapore and only wants a fair price.	

Source: authors' compilation from print media data.

Raw data summary

Singapore and Malaysian news

Methodology:

1.   The main points from the print media are listed here according to news source. The data presented below are direct quotes from print media clippings.
2.   All print media data presented here is linked to the Singapore–Malaysia water relations. News articles however that directly refer to the Singapore and Malaysia water relations are labelled in bold. Opinion, Editorial, Letters and Forum articles are also labelled accordingly to distinguish from news articles. The print media data in regular typeface are linked to the Singapore–Malaysia water relations and provides the context of the events.
3.   A positive [+], negative [-] and neutral [=] signage is then labelled according to the portrayal of bilateral relations. Only articles that pertain to Singapore and Malaysian water relations are labelled.
4.   The main points from print media articles in Malay and Mandarin are distinguished in italics. Due to limitations brought by translating these data into English, these are used as supplementary data. Further, these articles are not labelled according to positive, negative and neutral articles because only selected excerpts of these articles were translated to English.

Table 7A.6 Singapore and Malaysian news

Events	Singapore news	Malaysian news	International news
January 1993	• SG and Indonesia signs a joint agreement to develop water resources in Sumatra. The aim is to supply water to Singapore and to the Riau islands. (Cua, 1993)		
December 1993	• PUB will spend 1.4B on development projects. 66M will go to water projects. One-third of this will be for expanding of the water distribution network in SG, 23M is for building water and sludge treatment works in Johor while 11M will be for laying pipelines from Johor to the Johor straits. (Sreenivasan, 1993)		
March 1995	• The water conservation tax and water rates will rise sharply in the next few years to reflect water's true scarcity and the cost of developing new sources. The PUB will launch the 'Save Water' campaign in June. (BT, 14 March 1995)		
April 1995	• SG may tap into reserves to build desalination plants to meet rising demands for water. Desalination plants however are ten times more costly than existing water treatment processes. (Soh, 1995)		
October 1995	• SG gov't is concerned about industry's excessive water use. (Ang and Lee, 1995)		
December 1996	• PUB has been awarded first place by the IWSA-ASPAC for outstanding achievement on UFW. (BT, 4 December 1996)		

June 1997	• An MP encourages SG to invest in desalination plants to ensure continued water supply after the expiry of the first water agreement. It can be affordable (BT, 4 June 1997) • MY and SG signed the Points of Agreement (POA) on Malayan Railway Land in SG in 1990. PM Dr M has new thoughts on this and to avoid legal difficulties, PM Goh suggested that they can use a wider cooperation framework including the sale of water to SG. (BT, 6 June 1997) (+) • SG can be self-sufficient in water but it will be costly, accdg to DPM Lee. PUB will also build a small reverse osmosis plant for pilot study. (Srinivasan, 1997)
March 1998	• PUB has made its final recommendations to the gov't to build the desalination plants for SG. Construction is set to start by middle of next year and by 2003 the first desalination plant should be completed. (Lim, 1998a)
April 1998	• MY and SG has almost reached an understanding on the water supply agreement. MY has agreed to continue supplying water to SG beyond the current water agreements that expire in 2011 and 2061, as well as the 1990 agreement. MY also wants the specific terms to be made closer to expiry date because MY can't anticipate the situation at the time when the present agreement expires. The terms and conditions

Table 7A.6 (Continued)

Events	Singapore news	Malaysian news	International news
		will be mutually agreed within 60 days. (Abdullah and Bingkasan, 1998) (+) • MY may sell treated, instead of raw water, to SG (Kandiah, 1998) (+) • MY and SG has reached an understanding over water supply after 2061 but no agreement has been signed. (Abdullah, 1998a) (+)	
June 1998	• SG will continue with water conservation efforts and raise water tariffs and conservation tax as planned on 1 July. (Kau, 1998) (–) • SG and MY have not reached an agreement to extend water supply to SG after the expiry of current agreements. (Ong, 1998)	• Pahang government is keen to supply water to SG but will leave the decision to NWC. (Aziz and Haron, 1998) (+)	
July 1998		• PM Mahathir says MY and SG could not reach an agreement on the details of the water supply agreement. (Abdullah and Rajendram, 1998) (–) • Opinion: MY should terminate water agreement with SG because water pact just does not add up. (Wong, 1998a) (–)	• Disputes taking toll on ties between Singapore, Malaysia; clashes over water supply, pension-plan regulations spark war of words (Ng and Oon, 1998) (–) • PM Mahathir says MY has agreed to supply

- water to Singapore but with conditions. In principle, an agreement has been reached but the two countries have not reached an agreement on the details. (DPA, 7 July 1998) (+)
- PM Mahathir says that MY has agreed in principle to supply water to Singapore but there would be conditions imposed (BBC, 13 July 1998) (+)

- MY threatens to cut Singapore's water supply (BP, 4 August 1998) (–)
- MY leader fires hot words at SG; ties strained over water and a checkpoint (Fuller, 1998) (–)

- Umno youth urges MY gov't to be firm in defending its position on issues related to bilateral ties. They feel that SG FAM Jayakumar and HAM Wong did not reflect the republic's sincere stance. The Umno youth hoped that the government will fix a more realistic period and reasonable price for the water supply agreements (Haron, 1998) (–)
- PM Dr M assures that MY will not cut water supply to SG. (NST, 5 August 1998) (+)

August 1998

- Umno youth calls for the suspension of fresh dealings with SG and for a review of water supply agreements when they expire in 2011 and 2061. According to Umno youth, SG did not reflect a 'sincere stand'. MY should fix a time frame for water supply and realistic rates. The next day MY acting FM denies that it has decided on a freeze in ties. (BT, 4 August 1998) (–)
- SG is still keen to buy MY water. (Boey, 1998)
- American water treatment company, US Filter, opens its regional HQ in SG (BT, 5 August 1998) (+)
- PUB embarks on desalination, tapping existing water sources in MY and SG, tapping alternative sources in Indonesia for its long-term water supply. (Lim, 1998a)

Table 7A.6 (Continued)

Events	Singapore news	Malaysian news	International news
September 1998		• PM Dr M said that SG used the weakness of its neighbours to achieve economic prosperity. He did not see how MY could use the water issue to bully SG as Lee's memoirs claim. Dr M says SG was making more money than MY by selling water (Abdullah, 1998c) (−)	• PM Mahathir criticizes Singapore for rising 'old issues'. He was commenting on media reports on SG SM LKY's memoirs serialized in The Straits Times (BBC, 16 September 1998) (−)
November 1998	• MY has sought help from SG to raise funds to tackle the country's economic crisis, PM Dr M said. MY will also determine how it can meet SG's water needs. My agrees to continue supplying water to SG beyond 2061, that was to be confirmed within 60 days. (Wong 1998a) (+) • Speculation that MY is now inclined to link SG's proposal of financial assistance with the water deal (Kassim, 1998a) (=) • MY has sent back SG the draft loan agreement, it originally sent last 14 November. PM Goh said MY sought help from SG. Dr M reaffirms MY's commitment to provide water to SG. (Teo, 1998) (+)		• SG and MY hammering out water, funding deals; SG would help raise funds for MY's crisis-stricken economy and continue to receive water from its northern neighbour under deals now being negotiated, according to PM Goh (DPA, 23 November 1998) (+) • MY and SG agree to sink differences and work towards improving ties, which

have soured in recent months (TI, 6 November 1998) (+)

December 1998	• MY is not taking the US$4B loan facility from SG. Dr M proposed to link the issue of water to negotiations on all other bilateral problems: Malayan Railway land, relocation of Immigration checkpoint, use of Malaysian airspace by SG, transfer of MY shares on CLOB international to KL stock exchange, and early release of CPF to MY workers. (Kassim, 1998b) (=)	• MY has told SG that it does not need the RM15.2billion offered by SG as funds-for-water arrangement. MY has proposed all outstanding issues that have put a strain on bilateral issues be discussed as a package. (Abdullah, 1998b) (–)
January 1999	• Water deal with Indonesia is possible. (Sim, 1999) • Johor MB Ghani says that SG should continue to get its water supply from MY because it will be cheaper than buying from Indonesia. SG should also consider long-standing bilateral relations between the two countries before deciding on where to buy water says the Johor MB (ST, 21 January 1999) (+)	
February 1999		• Opinion: SG did extremely well despite shortcomings by 'wearing seven-league boots and spinning a lot of hype'. MY is short-changed in the waters supply agreement. SG created problems and it's up to SG to offer solutions. (Chin, 1999) (–)

Table 7A.6 (Continued)

Events	Singapore news	Malaysian news	International news
March 1999	• Forum: MY can choose not to buy water from SG. It is MY earning from the water SG sells. It is also MY who keeps changing its mind about the CIQ relocation. (ST, 3 March 1999) (–)		
June 1999	• SG will start a desalination programme. It is scheduled for completion by 2005. Florida method of reverse osmosis may be cheaper. (Lee, 1999) • SG ministry spokesman takes MY officials to task for having leaked details of 'confidential negotiations still in progress'. SG proposal for future water is always contingent on MY water needs. MY has insisted on negotiations from scratch and made new demands. (ST, 8 June 1999) (–) • Editorial & Opinion: SG has made it clear that water demand is always 'contingent to MY satisfying its own needs first'. MY is profiting from the treated water it buys from SG bec it sells it at RM3.95 per 1000 gallons when it buys it for RM0.5 from SG. Nevertheless, SG Must get away from relying solely on MY. (BT, 11 June 1999) (–) • SG will not go thirsty because it is always looking for more sources of water. (Leong, 1999)	• Johor will safeguards the interests of its people and their water requirements before committing to supply large quantities of water to SG. (Venudran, 1999) (=) • MY will supply SG with treated water after 2061. PM Dr M says the present agreement was drawn by the British and that it favoured SG at the expense of Johor (Singh and Sennyah, 1999) (–) • Johor must secure local water needs as a priority. SG has been making money from water drawn from Johor and they've been very demanding at current negotiations asking for a 3.4B L of daily supply after 2061. Johor however must secure its local needs first. (NST, 8 June 1999) (–) • Opinion: Johor also benefits from water agreements. They buy 37 mgd of treated water when SG is	• MY PM discusses package of issues between MY and SG. (BBC, 7 June 1999) (+) • MY PM discusses package of issues between MY and SG that includes water supply and the relocation of MY customs and Immigration facilities (BBC, 9 June 1999) (+) • SG refutes MY water profiteering claim and blamed KL for reneging on an agreement reached by the two leaders last year. (DPA, 9 June 1999) (–) • SG is looking for more sources of water

	and may turn to Indonesia after MY said it must give priority to its own needs before hiking the supply to SG. (DPA, 14 June 1999) (−)
	only required to sell 15 mgd. SG's request for water is always contingent on MY satisfying its own water needs first. (Tan, 1999) (−)
October 1999	• Pahang is keen to sell water to Singapore. (ST, 1 October 1999) (+)
January 2000	• AquaGen International will desalinate seawater with new technology to lower cost. (Nathan, 1999)
March 2000	• MY FM Syed Hamid responds to SG FM Jayakumar's comment and denies Umno elections are holding up talks with SG. Syed Hamid says it is the SG gov't that has not acted upon previous proposals made to them. Syed Hamid says it is pointless to have a meeting if there is nothing to be discussed since SG has refused to accommodate MY needs. (NST, 13 March 2000) (−)
April 2000	• The falling cost of desalination has made the production of good drinking water affordable. AquaGen is planning to desalinate water using the reverse osmosis method. (Khalik, 2000)

Table 7A.6 (Continued)

Events	Singapore news	Malaysian news	International news
June 2000	• A preliminary feasibility study has found that physical conditions in Riau are favorable to water resource development. PM Goh discussed with Pres. Wahid about the possibility of SG buying water from Indonesia on a commercial basis. (Kagda, 2000) • Stanford and NTU join up to look for ways to help SG produce water more cheaply and become self-reliant. (Tee, 2000)		
July 2000	• SG-Indonesia water project is floated. An estimated US$1.5B will be needed for the water catchment area in Riau. (Yeoh and Liang, 2000) • Indonesia as a source of water supply is being explored. (ST, 2 July 2000) • The new DTSS will improve water quality in the straits of Johor. (ST, 9 July 2000)		
August 2000	• MY government allocated RM 700M for the construction of a water treatment plant in Johor. This may introduce a new factor to the stalled bilateral talks. (Toh, 2000) (=)		
September 2000	• Domestic politics in MY has a lot to do with MY decision to build a new RM700M plant in Johor and stop buying from SG. (ST, 18 September 2000) (=)		
November 2000	• Gus Dur or Pre. Wahid of Indonesia says that SG would have no water if both MY and IND stopped supplies to the island. (Pereira, 2000) (−)		

December 2000	• Wahidologist think they've found an explanation to Pres. Wahid's outburst. He was visited by MY FM both times he made a public outburst. (Sim, 2000) (−)	• Johor will stop buying treated water from SG in 2003 because it will build a water treatment plant to be completed by 2002. (Chong, 2000) (=)	• SG minister ruled out the possibility of water cut-off from MY and stressed the need to diversify water sources for SG (Xinhua, 12 January 2001) (+)
January 2001	• Forum: SG must end reliance on MY and IND for water. (Ho, 2000) (−) • Beijing looks at SG for water conservation tips. (Hsien, 2000). • Hyflux opens a S$4M R&D center in Changi. (BT, 20 January 2001) • Drilling for the Deep Tunnel Sewerage System (DTSS) started yesterday. This reduces land requirement for traditional systems by 90 per cent (Liew, 2001)	• Opinion/Letters: SG is profiting from current water deals. Johor's move to be self sufficient is thus good (NST, 10 January 2001) (−) • Opinion/Letters: SG is earning from current water deals. The new agreements must be checked so that existing discontent is not blown out of proportion. (Maharis, 2001) (−)	
February 2001	• Indonesia and SG proceed talks on water supply to SG. (Kagda, 2001) • Riau-SG ties will include more than water. Riau stands to gain in FDI from this relations. (Pereira, 2001)		
March 2001	• It will soon be cheaper to desalinate sea water than to import it. (Kaur, 2001a) • SG is targeting to process 30mgd of desalination capacity for its first desal plant. NEWater is also being produced in a pilot plant in Bedok. (Kaur, 2001b)		

Table 7A.6 (Continued)

Events	Singapore news	Malaysian news	International news
April 2001	• Johor is aiming to be the second most efficient water distributor in MY. (Toh, 2001a) • SG opens a consulate in Riau (ST, 19 April 2001)		
May 2001	• Reclaimed water or NEWater will meet 20 per cent of SG's water needs. It will have a higher grade of purity than normal, potable water. (Low, 2001) • Groundbreaking of Changi water reclamation plant. SG is taking steps to ensure water supply. (Kaur, 2001c)		• Singapore is turning to technology to alleviate its water shortage. (NW, 16 July 2001)
September 2001	• SM LKY and PM Dr M have reached in principle agreement to resolve long standing bilateral issues. MY guarantees supply after the 1961 and 1962 agreements. A bridge and a tunnel will be built to replace the causeway, CPF will be returned over two years, etc. (Toh, 2001b) (+) • SM Lee visits MY PM and a skeletal agreement was reached. (Ng, 2001a) (+) • Worker's Party welcomes the recent agreements between SG and MY but is concerned about new water charges. (ST, 7 September 2001) (+) • SM LKY admitted that he conceded to PM Dr M because negotiations can become very protracted if there are major political changes in MY. (Toh, 2001c) (=) • Forum: SG should consider other sources of water supply because we can't rely on MY forever. (Tan, 2001) (−) • Commentary Analysis: The pact now hinges on details. (Ng, 2001b) (=)		• Summit yields significant progress. SG and MY have gotten together to set in motion efforts to resolve deadlock issues. (Oon, 2001) (+) • SG-MY agreement marks a psychological breakthrough says SG FM Jayakumar. (Xinhua, 25 September 2001) (+)

- Response to *Forum*: PUB is tapping all possible water sources. (Chan, 2001) (=)

October 2001

- Lt. Col. Azmy Yahya writes that MY must take full use of its advantage on water to leverage over SG and reduce incentive to use its military superiority to threaten MY. MY can 'pollute the supply with either chemical or biological agents' if SG uses its military strength. (ST, 9 October 2001) (–)

January 2002

- Johor chose not to review the water price in 1986 and 1987. The water agreements were also signed by the City council of SG and the Johor State Gov't. Thus the British could not play a role in an independent and sovereign MY. (Teh, 2002) (=)
- MY will name water price before talks start. (ST, 26 January) (=)
- SG subsidizes treated water sold to Johor. It gets no profits instead it subsidizes RM1.9 for every 1000 gallons sold to Johor. (ST, 27 January 2002) (–)
- SG denies unfair deal in water. SG asks MY to set out a 'clear framework' on the package of issues. British can't play a role in the existing deal because it was signed by MY Johor Gov't. (Chua, 2002a) (–)

- Discussions have stalled because MY and SG have failed to agree on the sale of water to SG. PM Dr M says SG would not settle for anything else before the price of water is determined. (Said, 2002) (–)
- SG has turned the water issue into a pivot for the package deal but it is holding the whole agenda hostage to the water issue risking sour relations. (NST, 23 January 2002) (–)
- DPM Badawi believes that the water issue must be resolved before there is progress on other issues. (NST, 26 January 2002) (=)
- MY is just as eager to solve bilateral issues. MY loses money daily with the delay in fixing the price of water. MY would not allow the

- MY & SG unable to agree on water supply price. (BBC, 22 January 2002) (–)
- SG denies MY allegations that water agreements favor the republic and denied profiteering from selling water bought from Johor. (BBC, 28 January 2002) (–)

Table 7A.6 (Continued)

Events	Singapore news	Malaysian news	International news
		issue to sour relations with SG. (Cheah, 2002a) (−) • Johor will no longer buy water from SG with the building of its treatment plant. (NST, 31 January 2002) (=)	
February 2002	• MY wants goods relations with SG but will insist for a higher water price. (BT, 1 February 2002) (+) • HK pays a higher price for raw water from Guangdong because mainland authorities have invested billions since the 1960s to build the infrastructure to deliver the water to HK. SG on the other hand has absorbed all costs of building the infrastructure costs in Johor. (Tan, 2002a) (=) • MY wants to maintain good ties with SG but this will not be at the expense of losing out in a new water supply agreement. MY only wants a reasonable payment for new deal. (Lau, 2002a) (+) • It is the sanctity of the agreement that is at stake. (Han, 2002) (−)		
March 2002	• Johor Gov't will submit a detailed report to SG on the land reclamation efforts. (BT, 4 March 2002) • PM Dr M sent a new water supply proposal to SG SM LKY. SG will study the implications of the new formula. (Toh, 2002a) (=)	• MY seeks assurance from SG that the latter's reclamation activities will not affect MY's deep water line. (Ahmad and Latiff, 2002; Sooi, 2002b) • Reclamation is affecting Johor ports. Mainline vessels have near	• SG studies new MY proposals in water price row. The basket of unresolved issues include the customs, immigration, and quarantine facilities

		misses with barges carrying sand to reclamation works. (Oor-jitham, 2002) • SG has asked MY to send a formal protest note if it feels that the land reclamation activities in SG have affected the deep water line (Cheah, 2002b)	• Opinion: the supply of water is immutable under the current water agreements but not the price or mechanism for it. A new price structure could not affect SG's sovereignty. MY does not like bilateral tensions, MY realizes the need for realism in foreign policy and diplomacy. (NST, 7 April 2002) (–) • Former Johor MB says SG once expressed willingness to go to war with MY if water is cut off. (Sayuthi, 2002) (–)	in SG, the redevelopment of MY railway land, use of MY airspace, release of CPF and the water agreement. (BBC, 12 March 2002) (=)
April 2002	• MY PM Dr M warned that MY would not cooperate with SG if it continued to reclaim land in the Johor straits. (BT, 12 April 2002) (–) • Opinion: Reclamation effort has taken the attention of PM Dr M who suggested that MY will not cooperate if SG did not reciprocate. This can be a proxy skirmish over the real issue-water supply to SG. (ST, 5 April 2002) (–) • PM Goh spoke about how MY water supply to SG has bedeviled ties for the past 37 years. The water agreements are internationally legally binding acts. A breach will call into question the Separation Agreement. (ST, 6 April 2002) (–) • PM Goh says the Singapore will rely less on MY for its water supply to prevent the long standing issue from continuing to strain ties between the two countries. (Tan, 2002a) (–) • Pahang wants KL to handle its water pact with SG. (ST, 30 April 2002) (=)		
May 2002	• SG expects NEWater to meet at least 15 per cent of SG water needs by 2012. This will supply water fabs with ultra pure water. (Tang, 2002)		

Table 7A.6 (Continued)

Events	Singapore news	Malaysian news	International news
	• MY can't simply alter its water-supply deal with SG it is bound by the terms of a pact that any change would require the consent of both countries. (Lau, 2002b) (–) • MP's raise concerns on MY-SG bilateral issues. (ST, 17 May 2002)		
June 2002	• Four taps will keep water flowing in SG. SG can be self-reliant if it wants to. (ST, 23 May 2002) • PM Goh expects good progress in bilateral talks with MY. He says it would be good to have a formula that can stand the test of time. (BT, 17 June 2002) (+) • SG and MY are expected to make good progress towards outstanding bilateral issues. However it is unlikely that they will solve all problems. (Lim, 2002b) (+)	• MY is prepared to take over water treatment plants in Johor if SG no longer needs to purchase water from MY. (NST, 22 June 2002) (–)	
July 2002	• SG and MY don't see a quick breakthrough on resolving bilateral issues but with the 'goodwill and the earnestness which exist on both sides to make progress', there will be progress. (Toh, 2002b; Chua, 2002b) (–) • MY has decided that it has the right to review the price of raw water sold to SG under the water agreements. MY has also unilaterally removed the negotiations on the current water agreements from the package of outstanding bilateral issues. Meanwhile SG FM Jayakumar says that water and the bridge project remain in the bilateral package. (Toh, 2002c; Chua, 2002c) (–) • MY has disclosed its plan to jack up the water price for SG by up to 100 times the existing prices. MY FM Syed Hamid said MY proposes 60sen/1000 gallons from	• MY is willing to supply water to SG for another 100 years, but the price will be revised. (SCD, 4 July 2002) • FM said that the water issue will be resolved if only SG is willing to release the CPF funds (UM, 5 July 2002) • Syed Hamid commented that SG agrees to allow Malaysians to withdraw their CPF savings amounting to 30 billion ringgit after the water issue is settled. (NSP, 6 July 2002)	• SG and MY FMs set for water and economic talks. (BBC, 1 July 2002) (=) • MY and SG agree to treat water issue separately. (BBC, 3 July 2002) (=) • MY and SG talks have failed to resolve the issue of treated and raw water prices. The four other

2002–2007, RM3/1000 gallons from 2007 to 2011. (Toh, 2002d; Ahmad, 2002a) (–)
- Experts find reclaimed water safe to drink. International panel gives SG two thumbs up after a two-year study. (Nathan, 2002b)
- NEWater gets clearance from a panel of international experts that it is fit for consumption. (Chuang, 2002a, 2002b; BT, 26 July 2002)
- NEWater can replace Johor supply. DPM Lee says water brought elsewhere must be competitive with reclaimed water. (Teo, 2002) (=)
- SG wants water price pegged to NEWater. (Chuang, 2002c; Lim, 2002c) (=)
- Jayakumar updates MP on three aspects of water issue: existing agreements, new agreements and price review. (ST, 24 July 2002) (=)
- Eight in ten SGeans prefer SG water as long as it is clean and fit to drink and support SG's move to become more self-sufficient. (Kaur and Hussain, 2002)
- Water will no longer be a strategic vulnerability of SG, accdg to FM Jayakumar. (Latif, 2002)

August 2002

- Views: If SG desecuritizes water, could it end 'hydro-politics'? (Kassim, 2002) (–)
- Mahathir is ready to end water pact. SG need not wait until 2011 to terminate one of its two water agreements with MY. MY did not want to sell to SG because it is a losing proposition for his country. (Pereira, 2002a) (–)
- Johor leader pokes fun at NEWater, warning MY that they risk drinking recycled water when in SG. (ST, 11 August 2002)

- Syed Hamid says that MY will not compromise on review of the water rate in the 1961 and 1962 agreements at the end of 25 years. (BH, 6 July 2002)
- FM Syed Albar says that the water issue would be the basis for the other outstanding issues to be resolved with SG. (BH and UM, 9 July 2002)
- FM Syed Albar says that MY is waiting for SG to settle the water issue first. (NSP, 9 July 2002)
- MY has urged SG not to delay in negotiation and solve the water pricing asap. (NSP, 22 July 2002)
- Our neighbour will be self-sufficient of their water needs after 2011. This will reduce tension between two neighbours. (NSP, 25 July 2002)

- SG must be reasonable about the price it pays to buy water from MY which does not make sense. (Bernama, 6 August 2002) (–)
- SG plans to let first of two water deals lapse in 2011. (NST, 6, 7 August 2002) (–)
- SG is free to stop buying water from MY if it wishes to accdg to PM Dr M. MY's continued

bilateral matters have been resolved. MY has stated its intention to exercise its right to revise the water rate. (BBC, 5 July 2002) (–)

Table 7A.6 (Continued)

Events	Singapore news	Malaysian news	International news
		supply of raw water to SG is more of a favour than trade. MY is interested in exercising its right to review the water prices. (Koh, 2002) (–) • PM Mahathir: It's fine if water pact with SG is not renewed. (The Star, 7 August 2002) (–) • Letters: SG propaganda is to make MY believe that it does not need to depend on MY for its water needs. It was MY not SG which said that after 2011 SG will no longer receive water under the agreement when it expires in 2011. (Omar, 2002) (–) • Johor MB Gahani said SGs announcement of allowing first of water agreements expire is not new. MY has already informed SG that all PUB land in Johor will be taken over by MY. Consumer Affairs and Domestic Trade Minister Tan says this is just SG's tactic to lure MY to agree to its proposed price mechanism. (Heng and Hamsawi, 2002) (–) • Editorial: It will be a thorn out of the flesh of bilateral relations if	

SG 2011 agreement expires. This should spell a clearer road ahead for bilateral relations. (NST, 8 August 2002) (=)

- Johor to stop buying treated water from Singapore once the Semanggar water treatment plant commences operations. (The Star, 8 August 2002a) (=)
- Flood of jokes over Newater. (The Star, 8 August 2002b) (–)
- Opinion: using newater as a bargaining ploy will not work with MY. (Tan, 2002)(–)
- NEWater, SG's newly discovered water source, is not expected to have a significant effect on MY's stand on the water pricing issue. (The Star, 18 August 2002) (=)
- NEWater will ease the tension of negotiation. (NSP, 2 August 2002)
- SG PM says that water supply must be based on mutual agreement, so as to maintain good relationship between countries. (SCD, 2 August 2002)
- SG PM says that SG will still buy water from MY, even though NEWater is enough to supplement SG's water supply. (NSP, 2 August 2002)

Table 7A.6 (Continued)

Events	Singapore news	Malaysian news	International news
		• MY not keen to extend the 2061 agreement said Datuk Syed Albar. (UM, 7 August 2002) • No assurance by FM Datuk Albar that water issue will be resolved at the 2 and 3 Sept meeting. (BH, 7 August 2002) • SG's decision to lapse the 2011 agreement was anticipated. (UM, 8 August 2002) • SGeans not comfortable accepting NEWater. (UM, 8 August 2002)	
September 2002	• Delinking the water issue does not make sense politically or tactically. (Chua, 2002d) (–) • Talks hit a snag after MY switched tack on the agreement to discuss the long-standing issues as a package. (Tan, 2002b) (–) • New efforts to collect rainwater from up to 2/3 of the island. (Kaur, 2002) • MY media takes potshots at NEWater. Columnists are taking a mocking tone of 'our water is best'. (Ahmad, 2002) (–) • MY will resort to arbitration to settle the water issue with SG. PM Dr M says there should a time frame in resolving the dispute over the price of water. (Lau, 2002d) (–)	• MY has expressed willingness to supply water for another 100 yrs at a negotiated price. SG and MY could still not agree on the pricing structure. (Osman, 2002a) (=) • MY has submitted a new water pricing structure. Syed Hamid said negotiations did not make much headway as both parties did not agree on the package of issues. MY proposed that only two issues – water price review and road and railway bridges to be discussed separately. MY	

would refer the matter to an independent arbitrator if talks failed. (Osman, 2002b) (–)

- SG has consented to discuss the water issue. MY says it has the right to review price. SG disagrees but is still willing to discuss the upward review. Meanwhile, Jayakumar says the talks on water should not be separated from the package of issues. (Osman, 2002c) (–)

- PM Dr M said there should be a time frame in resolving dispute over the price of water supplied to SG. Dr M says that it seemed SG did not want to accept the reality that the current water price is too low. Dr M says if SG wants to have water linked with package, MY will have no problems but it would prefer to deal with it separately because SG has not been reasonable in terms of water pricing. (Said and Loh, 2002) (–)

- SG's stand on water issue is a stumbling block in solving bilateral issues. (Chong, 2002) (–)

- SG DPM hopes that the second round of talks will see some progress. (SCD, 2 September 2002)

Table 7A.6 (Continued)

Events	Singapore news	Malaysian news	International news
		• SG needs to be sincere with MY. (Kamaruddin, 2002) • SG's ego is a hurdle in concluding the negotiation. (BH, 5 September 2002) (−) • The Permanent Court of Arbitration is ready to arbitrate water pricing problem. (BH, 19 September 2002)	
October 2002	• Johor MP suggested that MY sells SG sewage instead if it keeps insisting on negotiating down the price of water. (ST, 3 October 2002) (−) • PM Goh says that SG is willing to review water price if it is part of the package. PM Goh says that SG is not against an upward revision of the current price but that there must be a basis as to what the new price should be. (Toh, 2002e) (=) • KL presses SG to resolve water issue first. Other issues can be discussed later. NST and Berita Harian blamed SG for the slow progress in resolving the water issue. (ST, 11 October 2002) (−) • MY says it wants to discontinue the package approach on outstanding issues. PM Dr M also wants to backdate the price of water. (Low, 2002) • KL no longer wants to settle issues as a package. (Pereira and Lim, 2002) (−) • MY is giving SG one more chance in finding a solution to the water pricing issue, else it will be decided through	• SG has agreed to review the price of water, says FM Syed Hamid. PM Goh has indicated this to Dr M today. (Megan, 2002) (+) • SG wants to know basis for new prices of water. (The Star, 9 October 2002) (=) • Johor water may be sold to other states id SG decides not to buy from MY. (Begum, 2002) (−) • PM Mahathir said MY will backdate the new water price from the date it was supposed to revise the water rates for SG to recover the losses caused by the delay in the fixing of the new rates. (Bernama, 11 October 2002) (−) • PM Dr M says MY would recover	

- legal process. (BT, 14 October 2002; Ahmad, 2002b) (–)
- MY is stepping up the pressure on SG ahead of the negotiations as gleaned by remarks from senior officials and newspaper editorials. (Ahmad, 2002c) (–)
- SG tells MY that it has lost the legal right to review the water price. SG has shown flexibility in wanting to solve the issue. (Chuang, 2002d) (–)
- Concessions that SG was willing to make to secure a new water deal are now odd because MY has discarded the package approach. (ST, 16 October 2002) (–)
- Bilateral discussion must include both current and future water. MY agrees to this. (Toh, 2002f) (–)
- SG maintains that MY has no legal right to review the price of water. SG is sincere about wanting to make a progress in this issue. SG did not get adequate clarification to MY's explanation of how their proposed formula is consistent with provisions in the existing water agreements, and why they had not lost their right of review. (Chuang, 2002e) (=-)
- MY PM Dr M says SG was not showing any desire to reach a compromise in the water dispute with its 12 sens offer. (ST, 23 October 2002) (–)
- MY wants a higher price for the water it sells. MY PM says SG did not cut its offer price to 12 sen as PM Dr M claimed. SG clarifies that it wants to buy water from MY if the price is right. The 12 sen is SG's calculation of what the revised price of raw water would be if they proceed strictly on the basis of the factors in the review clauses. (Tan, 2002b) (–)
- MY is now mulling over the possibility of enacting laws to safeguards its own water needs and possibly supersede losses from past water agreements with SG by having price backdated. He says SG has been getting cheap water from MY. Dr M says PM Goh agreed to discuss the talks separately but now he doesn't. 'It is difficult to negotiate like this'. (Said and Darshni, 2002) (–)
- MY will seek to backdate the price of water sold to SG when both countries finally come to terms on the new rates, PM Mahathir said. (Leong, 2002) (–)
- MY may seek legal recourse if no headway is made in the next round of talks, says FM syed Hamid. Syed Hamid appeared irked when SG said Dr M misunderstood PM Goh. The upcoming meeting will be on water alone because SG asked for it. 'They agreed to that. In fact, they asked for it. Water has to be discussed first, the other issues must come later'. (NST, 14 October 2002) (–)
- Be sincere in resolving water price deadlock, S'pore told. (Lam, 2002) (–)
- MY says issues will be tackled separately. SG foreign ministry

Table 7A.6 (Continued)

Events	Singapore news	Malaysian news	International news
	the two water agreements with SG. (Ahmad, 2002d) (=)	says that trade offs are no longer possible if the package approach is discontinued. (Cruez, 2002) (–)	
	• SG said that comments from MY PM of considering a new law to render existing agreements null and void – are not in accord with repeated assurances from the MY gov't that agreements will be honored. (Chua, 2002e) (–)	• SG is reluctant to solve the water issue. SG is not keen to solve this because they will continue to get 3 sen per 1000 gallons. (NST, 24 October 2002) (–)	
	• Any attempt to breach the water agreements will only strain bilateral relations. (Toh, 2002h) (–)	• SG prepared to discuss new formula for water pricing with MY. (UM, 9 October 2002)	
		• Discussion on water issue did not reach an agreement. (Jafar, 2002a)	
		• SG claims that MY did not provide adequate information. (UM, 19 October 2002)	
		• In a recent meting, it was noted that MY's stand on the water issue, is more tolerant compared to SG. (Jaafar, 2002b)	
		• Spokesman from SG commented that MY was not sincere in resolving the water issue. (SCD, 19 October 2002)	
		• SG FM officials commented that SG is sincere in resolving the water issue. (NSP, 19 October 2002a)	

- DPM said that both countries need to be patient in resolving the water issue. (NSP, 19 October 2002b)
- DPM Badawi commented that SG's attitude which does not want to accept MY's proposal has caused the water talks to reach a deadlock. (BH, 19 October 2002)
- Johor minister says that SG must make a firm stand on resolving the water issue. (NST, 21 October 2002; Mustafa, 2002; SCD, 21 October 2002)
- MY will not use water as a weapon against SG. (SCD, 22 October 2002)
- MY will not use unresolved water issue was a threat against SG. (NSP, 22 October 2002)
- Price of water should exceed 60 cents for every 1,000 gallons. (NSP, 24 October 2002)
- MY can develop new act to resolve water issue. (BH, 25 October 2002a)
- MY has its rights to terminate the water agreement. (Arof, 2002)
- MP for Sri Gading Datuk says that MY should increase the water price, to teach SG a lesson. (BH, 25 October 2002b)

Table 7A.6 (Continued)

Events	Singapore news	Malaysian news	International news
		• Datuk rais Yatim, informed that MY has its rights to restrict the export of water to SG when the act is passed. (SCD, 25 October 2002a)	
		• Dy FM says that MY has the right to review the price of water. (SCD, 25 October 2002b; NSP, 25 October 2002a)	
		• MY to pass new act to regulate water supply. (NSP, 25 October 2002b)	
		• Former FM of MY says that the Water agreement should be terminated and MY should be able to back-up their decision. (UM, 26 October 2002)	
		• PM Mahathir says that supply of water to SG should be studied from every aspect. (UM, 26 October 2002)	
		• PM says that MY should study the from the legal side of the water agreement on the profit that SG makes from the water supply. (SCD, 26 October 2002)	
		• PM says that MY will study from the legal aspect on the supply of water to SG. (NSP, 26 October 2002)	

November 2002

- 'Singapore is trying to sell-out the country', says chairman for Johor Farmer Association, Encik Ibrahim Atan. (Musa, 2002a) MY Islamic Youth Organization wants the water issue to be resolved immediately, says its President En. Ahmad Azam Abdul Rahman. (BH, 26 October 2002)

- SG blames MY for breakdown on water talks. FM Jayakumar laid the SG case before parliament quoting from recent correspondence between the two countries' prime ministers to show that it had been willing to make concessions, but that MY was not prepared to compromise. (BBC, 1 November 2002) (–)

- SG will not stick to the letter of the 1961 and 1962 agreements in the ongoing water talks with MY. FM Jayakumar says, now that MY delinked water and unilaterally abandoned a package deal approach, all concessions are off. (Chuang, 2003g) (–)

- Johor had confirmed in 1990 the 3 sen per 1000 gallon price to Singapore when it signed an agreement with Singapore's PUB for the construction of the Linggiu Dam, says FM Jayakumar. (BT, 1 November 2002) (=)

- FM Jayakumar explains to parliament that MY held up talks. Quoting extensively from recent correspondence between the two countries, it shows that SG is willing to do concessions but MY was not prepared to compromise. (Lim, 2002d) (–)

- MY FM Syed Hamid says that the water dispute with Singapore may have to be resolved in court but that it was in MY's interest to find a solution. (BT, 2 November 2002) (–)

- MY have asked their gov't to let the two water pacts with SG lapse, adding that there was no need to

- Dr M defends MY right to price review because SG should not have suggested a price revision in the first place if MY didn't have this right. Dr M reiterates that MY will not change its stand to settle the water issue individually. (Abu Bakar and Chow, 2002) (–)

- SG FMinistry says that MY should be prepared to seriously discuss SG's water supply after the 1962 agreement. Then SG would be prepared to discuss the current price of water. (NST, 21 November 2002) (–)

- MY has decided to stop negotiations with SG and instead will seek legal recourse for price water review. (NST, 30 November 2002) (–)

Table 7A.6 (Continued)

Events	Singapore news	Malaysian news	International news
	convene talks again. There was even a suggestion that the gov't cut off water supply to SG first then proceed with talks with SG. (ST, 8 November 2002) (–)	• PM says that we have rights to review the water pricing. (UM, 2 November 2002)	
	• MY expects a new round of talks next year. The talks on the price of water will continue until both sides agree, accdg to Syed Hamid. (ST, 16 November 2002)	• MY has given SG the ultimatum whether it wants to continue negotiation or refer the matter to a third party says Syed Hamid. (Ibrahim, 2002)	
	• KL does not want to admit that it was MY, not SG, that got the ball rolling again. It was incorrectly interpreted that SG is softening its stance. (Pereira, 2002b) (+)	• MY will change its stand on the water issue says PM Mahathir. (BH, 2 November 2002)	
	• MY is only willing to discuss the current price of review water and not a new agreement. (Pereira, 2002c) (–)	• FM Syed Hamid says that SG needs to be sincere in their stand on the water issue. (BH, 2 November 2002)	
	• SG is restates its stand that current price and future supply should be discussed. (ST, 21 November 2002) (–)	• SG should be grateful to MY for agreeing to supply water to them and not adopting the delaying tactics. (UM, 4 November 2002)	
	• Little is expected of next round of water talks. Syed Hamid says the meeting will only discuss the price of water MY is now supplying to Singapore. It will not talk about future water unless Singapore accepts MY's right to a review. SG's foreign ministry says that little will be achieved if this is MY's position. (BT, 21 November 2002a) (–)	• People in Johor urge the government to be firm in dealing with SG. (BH, 4 November 2002)	
	• United Engineers has won a $200million contract from the PUB to carry out mechanical and electrical works at the Changi Water Reclamation Plant. (BT, 21 November 2002b) (–)	• Johor will take over the treatment plants at Gunong Pulai, Tebrau & Seudai from PUB when the agreement expires in 2011. (BH, 6 November 2002)	
		• SG saves RM1.5B from the sale of water supply from MY. (UM, 13 November 2002)	

- There should be an increase in the assessment rates for PUB plants operating in Johor. (UM, 14 November 2002)
- Higher taxes should be charged to PUB plants operating in Johor, says Kempas State Assemblyman Sapian. (Musa, 2002b)
- MY intends to continue negotiation with SG on the water issue, says Syed Hamid. (UM, 15 November 2002)
- SG and MY continue discussion on the water issue next year. (UM, 18 November 2002)
- SG puts forward two conditions on future discussions towards the negotiation on the water issue. (BH, 18 November 2002)
- SG is purposely delaying the discussion says Johor MB. (BH, 22 November 2002a)
- MY's stand is the cause for the delay, says SG FM. (BH, 22 November 2002b)
- Jayakumar says that MY press has distorted the message of PM Goh. (SCD, 2 November 2002a)
- PM says MY will study effects of implementing the new act to regulate the sale of water to SG. (NSP, 2 November 2002a)

Table 7A.6 (Continued)

Events	Singapore news	Malaysian news	International news
		• MY FM says SG is not sincere in resolving water issue. (NSP, 2 November 2002b; SCD, 2 November 2002b)	
		• Rais Yatim: cabinet will only consider court after all negotiations have failed. (BM, 3 November 2002a; SCD 3 November 2002)	
		• Syed Hamid said that he would propose a motion in Parliament next week to debate the water issue. (BM, 3 November 2002b)	
		• SG FM says that both should be sincere in solving the water pricing issue. The cost of treated water is RM2.5/ 1,000 gallons but SG is selling treated water to Johor at RM 0.5/1,000 gallons. (NSP, 14 November 2002)	
		• Higher taxes should be charged to PUB plants operating in Johor, says Kempas State Assemblyman Sapian. (SCD, 14 November 2002)	
		• MY is waiting for SG to fix the next round of meeting, says Syed Hamid. (NSP, 15 November 2002; SCD, 15 November 2002)	

- FM says that MY still wants to negotiate on the pricing of water. (SCD, 18 November 2002)
- Syed Hamid says MY will only discuss on water supply until 2061. SG FM says that there will be not much progress if the talks only discuss water pricing until 2061. (SCD, 21 November 2002)
- SG is purposely delaying the discussion says Johor MB. (NSP, 22 November 2002)

- In response to reports quoting SG FM Jayakumar saying SG will look into arbitration, FM Syed Hamid says there is no reason for the issue to be taken to arbitration. (NST, 30 December 2002) (–)
- MY wants to settle the outstanding water issue through legal means. (NSP, 2 December 2002; SCD, 2 December 2002a)
- SG has not received any formal notice for the next round of meeting. (SCD, 2 December 2002b)
- PM Mahathir has assured that water supply to SG will not be cut off. (SCD, 27 December 2002)

December 2002
- MY gov't has directed the FM's legal department to prepare measures for legal recourse. FM Syed Hamid told the ST that MY is calling off water talks with SG. (Lau, 2002e) (–)
- SG is waiting to receive official clarification from MY on future water talks. (ST, 1 December 2002) (=)
- KL refuses to confirm that it is seeking arbitration. (Ahmad, 2002e) (–)
- Syed Hamid says that SG must recognize MY right to review the price of water (ST, 11 December 2002) (–)
- Forum: SG would agree to a new price if it's reasonable. (Chuieng, 2002) (=)
- MY is serious about arbitration. (ST, 28 December 2002) (–)
- SG is ready to go for arbitration over water issue though it would still prefer to settle the matter through negotiations (Lim, 2002e) (–)
- KL will not go to court over water pricing but instead use local laws to settle the issue. (Ahmed, 2002f) (–)

Table 7A.6 (Continued)

Events	Singapore news	Malaysian news	International news
		• Dr Mahathir says if the water dispute cannot be resolved between both countries, it will be referred to a third party. (NSP, 27 December 2002)	
January 2003	• Johor will stop buying treated water from SG from the middle of this year with the completion of its water treatment plant in Semanggar. (BT, 7 January 2003) (=) • Hyflux in deal to draw water from air. (Boey, 2003a) • Johor MB Ghani offered the reason why SG gov't should agree to a price review of water. SG he says must have dignity. The MB says SG has a siege mentality, the more western their approach and tend to focus self interest in their relations with others, the more SG-MY relations are deteriorating. (Pereira, 2003a) (−) • Johor intentionally did not review prices in 1986 or 1987 because Johor was dependent on SG for treated water and SG would have increased its price if Johor had charged for more raw water. This is accdg to Johor Assembly Speaker Zainalabidin Zin. (ST, 20 January 2002) (=) • SG awards its first seawater plant tender to SingSpring. It will be operational in 2005. (Kaur, 2003a) • The water dispute is not about money but about SG's sovereignty and about honoring agreements. Jayakumar says that the vital issues is not 'how much we pay, but how any price revision is decided upon'. (Ibrahim, 2003) • SG releases letters. Jayakumar explains in the Parliament	• Letters: MY has been behaving like a good big brother to SG. MY has to get tough with SG. Author suggests to cut off water supply to SG for a start then perhaps SG will be interested in in their own health and leave MY in peace for a while. (NST, 2 January 2003) (−) • Johor's intention to stop buying treated water from SG has nothing to do with current dispute. Beginning the middle of 2003 with the completion of its water treatment plant, Johor will discontinue buying from SG. (NST, 9 January 2003) (=) • FM Syed Hamid questions SG's action in making public correspondence between the leaders of both countries and the diplomatic exchanges on negotiations for the price of water. 19 letters	• SG introduces NEWater to its tap. Relying on neighbouring MY for about half of its water supply, SG's move towards self-sufficiency seems to be prompted by pride. (TE, 11 January 2003) (=)

about the SG-MY water relations. (ST, 26 January 2003) (–)

- SG's stance on water gets scant coverage in KL papers. (Pereira, 2003b) (–)
- M'sia wants 200-fold hike in price of raw river water; Jump to RM6.25 per 1,000 gallons latest position shift. Any discussion of future water supply after 2061, when the longer of the two agreement expire, should take place only in 2058. (Chuang, 2003f) (–)
- PM Mahathir rules out war with SG and maintained that the heart of the water dispute is price and not sovereignty. He accused SG of the 'hate campaign' against MY to divert attention from its own serious internal problems. (Pereira, 2003c) (+)

were made public yesterday, today five more were published. (Said, 2003a) (–)

- SG is making MY a scapegoat to divert its citizens from its international problems, according to PM Dr M. (Mohd et al., 2003) (–)
- PM says that the MY government will try to resolve the outstanding water issue, without involving legal means. (NSP, 3 January 2003)
- The setting up of water treatment plant by the Johor State Government is not an unfriendly act and is not against the water supply agreement said FM Syed Hamid. (SCD, 8 January 2003)
- Johor does not need treatment water from SG anymore. The Semanggar water scheme has the capacity to treat 160M liters of water per day. (Nasir, 2003)
- Johor needs to treat its own water, says Johor MB. (Raaff, 2003)
- MCA Youth Johor says that SG is not matured. (UM, 28 January 2003)
- SG exposed the correspondence between both countries because they are pressurized to find a solution. (UM 29 January 2003)

Table 7A.6 (Continued)

Events	Singapore news	Malaysian news	International news
		• Johor Menteri Besar has informed the public not to get involved in the outstanding issues involving both the countries as both governments will resolve the differences. (SCD, 20 January 2003a) • Johor Menteri Besar informed that the state government would take back the water treatment plants after the 2011 agreement expires. (NSP, 20 January 2003) • Singapore Straits Times is quoted saying that Johor benefits most from the sale of water between both countries. (SCD, 21 January 2003) • The SG side of reporting of the water issue is misleading and does not reflect the true happenings. (SCD, 29 January 2003a) • SG should stop pointing fingers, says MY's top minister. (BH, 28 January 2003a) • Describing as a lack of good faith, FM MY questioned Singapore's action in making public correspondence between the leaders. (Said, 2003a)	

- MY will stop supplying water to SG when the last water supply agreement expires. (F. Abdullah, 2003)
- MY has rights to stop supplying water starting 2011. (BH, 29 January 2003)
- SG tries to divert the attention of its citizen's from economic problems. (UM, 30 January 2003)
- Once the agreement lapses, the supply of raw water to Singapore will stop. (NST, 30 January 2003)
- PM says that SG should not manipulate the issue and talk about going to war, when the actual issue is about the pricing of the sale of water. (NSP, 31 January 2003)

- SG is firm on the issue of sovereignty and is not twisting its facts. (BH, 1 February 2003)
- The SG Government reiterates tat the water dispute between the two neighbours is not about money. (NST, 1 February 2003; The Star, 1 February 2003)
- Disagreement with SG because of the release of diplomatic correspondence. (Moses, 2003)
- Opinion: MY has always agreed

February 2003

- SG: the real issue is not water price but the risk to SG's sovereignty if KL decides to change the agreements unilaterally. (ST, 1 February 2003) (–)
- Johor MB Ghani: the price has nothing to do with SG's sovereignty. (ST, 3 February 2003) (–)
- SG's move was an effort to divert attention. Its aim was to set the record straight. SG's clearly stated position is a reasonable one – it is not how much to pay for Johor water, but how any price revision is decided upon. One silver lining is Dr M's comments on MY being prepared to supply water after 2061. (BT, 6 February 2003) (–)
- SG and MY signed an agreement to refer Pedra Branca

Table 7A.6 (Continued)

Events	Singapore news	Malaysian news	International news
	to ICJ. Jayakumar suggest that a similar third party approach be used to resolve water talks. MY Syed Hamid says MY has not passed any law to stop its supply of water to SG. SG asks to respect status quo with Pedra Branca and MY says it has its own definition of the status quo. (Lim, 2003)	the sanctity of the water agreements. It is the price that is the issue now. (NST, 2 February 2003) (–)	
	• SG and MY sign pact to refer Pedra Branca dispute to ICJ; tensions likely to persist pending court's verdict. (Chuang and Toh, 2003a)	• Johor's MB says that the issue is not about sovereignty but about pricing. (SCD, 3 February 2003)	
	• MY PM expressed love for Singapore on Valentine's Day saying 'We love you. On Valentine's day, we love you.' 'Oh, people of Singapore . . . We are your friends, we do not want to have war with you, we cannot afford to fight with you' except if you go to war with us. If you step without our permission on Johor's soil, we are forced to slap you'. (BT, 15 February 2003) (-/=)	• SG FM says that the issue is about sovereignty and not pricing. (NSP, 3 February 2003)	
	• Singapore launches NEWater supply to reservoirs, water fabrication parks and commercial hubs. 2M gallons are being pumped each day to reservoirs, 10M by 2011 (Tan 2003)	• MY is unhappy with water pricing but will honour the agreements. Both foreign ministers agree that the water issue must be resolved in the spirit of goodwill as the dispute over Palau Batu Puteh is handled. (Hong and Abdullah, 2003) (=)	
	• SG–MY deadlock is more than just about water; its about the mutual fears of two closely intertwined neighbours. (BT, 26 February 2003) (=)	• Water: Singapore to be self-sufficient by 2061. (NST, 10 February 2003)	
		• MY water supply act will be reviewed says Syed Hamid. (UM, 10 February 2003)	
		• SG will have sufficient supply of water by 2061, says SG Environment Minister Lim Swee Say. (BH, 10 February 2003; NSP, 2 February 2003)	

- There should be avenues for amicable solutions which would set both countries in a path of cooperation and win-win dealings. Both nations should uphold the sanctity of the agreements. (Chia, 2003)
- SG denies that it created a deadlock by releasing confidential documents to the public. (NSP, 13 February 2003)
- SG must change its negative attitude which is not helping in solving the mutual problems. (UM, 20 February 2003)
- MY will supply water to SG until the end of time. (SCD, 20 February 2003f)
- MY has no intention to stop water supply to SG. (NSP, 20 February 2003)
- Singapore is urged to come down to earth and try to solve common problems. (NST, 20 February 2003)
- MY wants to continue the relation with MY and FM Jayakumar. (BH, 1 March 2003)
- If MY and SG cannot settle the water issue, MY will settle the case to world court said Yatim. (SCD, 1 March 2003)

March 2003

- NEWater is safe to drink. Over 22,000 tests done in two years. (ST, 1 March 2003)
- From Singapore Creek to Johor River; Old municipal records show that the British colonial administration have made water planning in Singapore a long-standing pre-occupation. Things have come a full circle. (Kassim, 2003b)

Table 7A.6 (Continued)

Events	Singapore news	Malaysian news	International news
	• Give engineers due credit for contributions. (Tse, 2003) • Overseas firms thirst for NEWater impressing American facilities in the business. (Kaur, 2003b) • Wringing water from air: the battle hots up; US firm Excel wades in with two types of machines. (Kassim, 2003c) • Gov't to keep water tariffs affordable. (Boey, 2003b) • Syed Hamid: MY is prepared to discuss the supply of treated water to SG. (ST, 26 March 2003) (+)	• SG published book to defend itself on the water issue. (BH, 17 March 2003) • Negotiation on water talks has reached a deadlock. MY will refer the case to ICJ. (SCD, 26 March 2003) • If the water talks fails, MY will refer the issue to the ICJ. (NSP, 26 March 2003a) • FM says that SG's accusation on MY on the failure of water talks as baseless. (NSP, 26 March 2003b)	
April 2003			• Singapore touts NEWater. (Miyauchi, 2003)
May 2003	• KL's Eco Water eyes Sesday Listing and has lodged its preliminary prospectus with the Monetary Authority of Singapore. (Lim, 2003a)		
June 2003	• Sinoemen launches 100m share IPO to raise $41.5M; Analysts see it as promising investment choice in water sector. (Lim, 2003b) • MY will honour its obligation to supply water to SG. Changes in control of water resources will not affect water deals. (Ahmad, 2003a) (+)		

- SG could become a hub for water treatment companies to raise funds, with more set to join those that have already listed in the country. (Buenas, 2003)

July 2003

- Syed Hamid: MY will produce its own book to give its version of the water dispute. (Lau, 2003c) (–)
- MY is confident that there is still room for talks with SG on the water issue. (ST, 8 July 2003) (+)
- KL launches ad blitz over water dispute; it claims SG made RM 622M from selling processed water to Malaysia. (Toh, 2003a) (–)
- KL ad blitz a rehash of old stories: MFA. SG appears unfazed by the latest newspapaer ad campaign that MY has launched against it over the two countries' water dispute. (Chuang and Toh, 2003b) (=)
- KL blitz ignores crucial facts, says SG. FMinistry says it is a rehash of old arguments and is puzzled by the timing of the current campaign against the republic. (Lee, 2003a) (–)
- KL has not closed door on talks. (ST, 20 July 2003) (+)
- PAS described that the media blitz is a waste of public money. (ST, 23 July 2003) (–)
- SG gets an ad in the Asian Wall Street Journal explaining its position. (ST, 26 July 2003) (–)
- SG responds with water ad; SG placed a full-page ad in the AWSJ yesterday to put the facts of the water dispute with MY and SG position – on the record. 'This is not the start of an ad campaign. The advertisement is to put the facts and SG's position on the record' (Toh, 2003b) (–)

- MY criticizes publication of letters bet SG and MY. (Lian, 2003) (–)
- MY releases advertisements in local newspapers on 'Water: The Singpaore-Malaysia Dispute: The Facts': SG earns PM 662M profit from the raw water supplied by MY. (Said, 2003b) (–)
- 'Is 3 sen/1000gallons a fair price?' SG will pay so much more if it used NEWater. (Megan and Chan, 2003) (–)
- 'Why not a fair price?' Each SG need only pay 26 SG cents for all the water SG takes from MY for an entire year. (Said, 2003c) (–)
- 'Has MY lost the right to ask for a fair price?' The water agreements stipulate after 25 years for price revision. Why did LKY negotiate for water price if MY lost its right? (Said and Cruez, 2003) (–)
- 'All points have never been agreed'. LKY saying that there is no agreement until all points are agreed and signed by the two PMs. (Said, 2003d) (–)

- MY PM defends public message on SG water issue. (BBC, 16 July 2003)
- MY issues booklet on water dispute with SG. NEAC published the 'Water: the Singapore-Malaysia Dispute: Facts' (BBC, 21 July 2003) (–)
- MY FM says that the door for negotiation is still open to SG towards resolving the water dispute. (BBC, 22 July 2003) (+)
- SG places advertisements on water dispute in five MY newspapers and received and instant rejoinder from KL in the very same editions in all five papers. (BBC, 29 July 2003) (–)

Table 7A.6 (Continued)

Events	Singapore news	Malaysian news	International news
		• MY is not mean and miserly in water deal. If it buys Newater, its costs would escalate. SG subsidized MY by RM25M while MY subsidized SG by RM478.4M. (NST, 18 July 2003a) (−)	
		• 'Is MY the big bully?' MY is not a big bully. MY has been subsidizing SG 18 times more than what SG claims it has been offering. (NST, 19 July 2003a) (−)	
		• Opinion: MY publication of facts on the water dispute signals MY's intention to tell the people and the world that its demands are just. Current water price is too low. (NST, 19 July 2003b)(−)	
		• The booklet, 'Water: The Singapore–Malaysia Dispute, The Facts' is sold at 3 sens a copy. This is to inform the public about the actual situation pertaining to the stalled discussion on the supply of water. (Abdullah, 2003a) (−)	
		• SG is depleting a fast evaporating reservoir of good will. SG reluctance to pay an affordable and fair price for water is puzzling. (Abdullah, 2003b) (−)	

- Letters: SG's offer of 45 sens per 1,000 gallons is an insult to MY (−). (Lee, 2003) (−)
- Letters: SG is so kiasu that it will never pay MY a fair price. We might as well give water free instead of wasting time and effort trying to resolve the issue. (Aziz, 2003) (−)
- MY responds to SG's ad in the Asian Wall Street Journal saying that it had no choice but to respond yet again to SG's claims and misrepresentations. (D. Loh, 2003) (−)
- SG always finds ways to hurt the feelings of its neighbours. (Pilihan, 2003)
- SG should be responsible for the media war. MCA Youth director Fan Lee Ee said SG should be fully responsible for the advertisement blitz over the water issue between MY and SG. (ODN, 16 July 2003)
- SG is urged to stop accusing on the water issue. (Bakar, 2003)
- Resolve the water issue in a harmonious manner. (BH, 21 July 2003)
- SG should not continue arguing on the water pricing and spoil the

Table 7A.6 (Continued)

Events	Singapore news	Malaysian news	International news
		relationship between both countries. (UM, 21 July 2003) • NGO says that SG should pay a reasonable price for water. (SCD, 24 July 2003) • SG continues to publish false information in the press on the water issue. (BH, 26 July 2003) • Syed Hamid says that the advertisement by SG in the AWSJ is misleading. (SCD, 26 July 2003a) • SG FM says that he does not understand comments made by MY FM on resuming talks will commence if SG changes their attitude. (SCD, 26 July 2003b) • NEAC has published false answers in their advertisement reports Singapore in the AWSJ. (BH, 28 July 2003) • NEAC says MY will point out the actual facts by replying to the latest advertisement by SG. (SCD, 28 July 2003)	
August 2003	• MY gov't move could herald new water laws; Federal gov't move to wrest control of water management from states in the MY Peninsula could be a precursor to the enhancement of a law that may affect SG. (Toh, 2003c) (−)	• The water supply system in Hong Kong and Kwangtung can be an example for MY and SG to follow. (NSP, 1 August 2003)	• MY PM says that the federal government's move to take over management of water

	• KL 'will supply water' despite price dispute; Dr M again blames SG for odd-looking bridge (Toh, 2003d); 'I think the period of talking and negotiating is now over. We want to have arbitration' says Dr M (–) • Syed Hamid: MY will honour current agreements but the time for talks is over. (Ahmad, 2003b) (–) • PAS: water dispute with SG is a diversion with Federal gov't attempt to take over management of water resources from state govt's. (ST, 5 August 2003) (–)	• No more negotiation. Water issue to be settled through court. (Jaafar, 2003) • The decision of the ICJ will be the final judgment in resolving the water dispute between both countries. (SCD, 3 August 2003) • SG will delay the water negotiation till 2011, the PM wants to expedite the arbitration process. (Alias and Said, 2003) • SG will attempt to delay to resolve the water issue says PM. (Abdullah, 2003; SCD, 4 August 2003)	supply from the states would not affect MY's water supply to SG. (Xinhua, 1 August 2003) (=)
September 2003		• Water issue progressed secretly without media coverage. (UM, 1 September 2003) • Johor need not depend on SG for water when the Semanggar plant is ready. (UM, 3 September 2003) • Syed Hamid clarified that there was no secret talk on water issue. Both parties are preparing to refer the issue to arbitration. (NSP, 3 September 2003)	
November 2003	• Singapore is fast becoming a centre for water treatment and environment stocks. The latest to join the cluster of such stocks is Asia Environment holdings. (H.Y. Loh, 2003)		

Table 7A.6 (Continued)

Events	Singapore news	Malaysian news	International news
December 2003	• Singapore has unveiled a new attraction – a high tech plant which makes sewer water potable. (BT, 6 December 2003)		
February 2004	• Asia environment doubles on debut; interest in water plays, attractive valuation lead to heavy trading. (Koh, 2003) • There is a new spirit in MY. The new PM is Abdullah Badawi. (BT, 7 February 2004) • MY PM Badawi opened the new treatment plant in Johor. This cut Johor's dependence on SG treated water. (Osman, 2004) (=)		
June 2004	• SG firm to dip into China water business; Environmental Holdings has tied up with Malaysia's Gaang in a new joint venture. (Toh, 2004)		
November–December 2004	• SG PUB launched WaterHub which aims to be a premier center for research and development projects. (Khin, 2004)		• SG aims to build itself into a global hydrohub by growing its industry share 3 to 5 per cent in the global water market. (Xinhua, 23 Novmber 2004) • MY and SG agreed to jump-start talks on unresolved bilateral issues with SM Goh

stressing that future relations should not be held hostage by past issues. (BBC, 14 December 2004) (+)

January 2005	• SG to increase water catchment area to two-thirds of the island. (Chua, 2005a)
Mar 2005	• Marina Barrage will create new reservoir to store 10 per cent of current water demand by 2007. (Ghani and Hooi, 2005)
July 2005	• Singapore will host a major conference on water and wastewater next month. (BT, 6 June 2005)
September 2005	• Water industry thriving in Singapore; around 30 firms working on NEWater projects worth $640M. (Huifen, 2005) • Gov't will invest $1.5B in water related projects. (Chua, 2005b) • PM Lee opens the Tuas Desalination plant. (Leong, 2005) • PUB expands to three NEWater facilities. (ST, 16 September 2005)
December 2005	• SG joins exclusive water research group. (Lee, 2005; BT, 7 December 2005) • UE bags PUB water treatment job. (BT, 21 December 2005) • UEL forms unit to seek regional water-treatment projects. (Phan, 2005)

Table 7A.6 (Continued)

Events	Singapore news	Malaysian news	International news
2006	• KL classifies confidential documents on SG issues; release of letters, meeting extracts attempts to counter Mahathir's claims that MY could have proceeded with the proposed bridge linking Johor to SG. (Ng, 2006) (=) • A new council, the Environment and Water Industry Development Council, is to be set up to spearhead the growth of the environment and water industry in Singapore. (Huifen, 2006) • The UN HDR recognizes SG as No. 1 in managing water resources. (Choong, 2006) • SG resource management can be a model for others: panel. Republic seen as global leader in successfully managing water. (Phan, 2006)	• With the water relations at an impasse, the plan to replace the causeway bridge might not work out. SG did not accept that MY had a right to unilaterally replace its side of the causeway with a half-bridge, it raised the fact that MY stopped SG from undertaking reclamation works within SG's territory in 2003. Any major work that relates to the Causeway, which carries pipelines supplying water to SG, would affect both countries. (NST, 14 April 2006) (−) • NST published excerpts of the four letters published by SG in 2003. (NST, 16 July 2006) (=)	• 'Water Queen' crowned for treatment business. (TNW, 22 May 2006) • Seeing Singapore through a drop of water. The Marina barrage and Marina reservoir are examples of how SG, despite being an island state with limited water resources, has turned this vulnerability into a strength. NEWater is another SG success story. (TKH, 9 August 2006) • Singapore taps ocean for water and income; Singapore now produces 30mgd of water daily from its desalination plant. (IHT, 12 September 2006)
2007	• Fourth NEWater plant will open next week. NEWater caters 7 per cent of the country's water demand. (Peh,	• Response to commentary on 21 May by Ridzam: the 1961 and	• Dr Mahathir urges renegotiation with

2007 (cont.)	• NEWater will meet 30 per cent of needs by 2011. (Tan, 2007)	• 1962 agreements are binding international agreements, not commercial contracts. The issue is not simply a matter of price, rather, on how the agreement on the new price is reached by both sides. (Nair, 2007) (=) • MY is keen to learn from SG's experience in river management and beautification. An MY delegation visited SG and toured Marina Barrage, SG River, NEWater Center, and HDBs. (Ramachandran, 2007) (+)	SG over water supply issue. (BBC,12 February 2007) (=) • SG is adding three reservoirs, moving to collect two-thirds of rainwater, up from the current one-half and promoting human reuse of treated wastewater. (WN, 29 October 2007)
2008	• SG holds the first Singapore International Water Week. (Mulchand, 2008) • Marina Barrage is completed. (ST, 10 July 2008)		• Singapore is positioning itself to become a global water hub. Investment opportunities are being offered by Singapore. (Calderon, 2008) • SG immersing itself in advanced water technologies. (Noma, 2008)
2009	• SIWW will draw 10,000 delegates. (Gunasingham, 2009) • Review: Persuade SGeans to drink tap water instead of bottled water whenever possible. (Koh and Leong, 2009)	• SG is a test laboratory that is becoming a major player in the water business. (Gomez, 2009)	Singapore offers a global model for sustainable solutions in water resource management. (BW, 7 August 2009)

8 Looking ahead

Introduction

The Singapore water story is one of consistent long-term and visionary planning followed by timely execution. It is one of leadership and relentless pursuit of ever-increasing economic growth, improvements of living conditions and quest for sustainable development. It is a success story of self-reliance and creation of growth opportunities, consistent with good governance, notwithstanding the scarcity of natural resources that would normally sustain such a development process.

Starting a mere five decades ago, the city-state's apparent vulnerability due to its small size, almost total lack of natural endowments and high water dependence on external sources has been turned into a myriad of opportunities propelling it and its population to continuously higher levels of development. The characteristic efficient, pragmatic and top-down decision-making in which policies have been formulated and implemented have made national strategic planning a distinctive and flexible process. Forward-looking and holistic thinking has been consistently channelled towards initiating required changes and anticipating and responding to new problems and challenges in what seems to be a permanent search for feasible and cost-effective opportunities.

During the early period of post-independence, Singapore focused primarily on enhancing its indigenous capacity, cleaning its water bodies and constructing and expanding catchments areas and reservoirs within an overall framework of sustainable land use. This in turn shaped urban development patterns and facilitated land and water conservation as well as strict implementation of regulations for storm and inland water management. Figure 8.1 shows the Blue Map of Singapore which indicates the catchment areas in 2011 making evident the water-related progress achieved from the time of the independence in 1965.

With time, water resources policies have evolved, aiming to bring water issues closer to the people in an attempt to conserve the scarce resource, promote social cohesiveness and sense of belonging and to create an environmentally attractive city where green landscapes and clean water are dominant features. This has improved the quality of life of millions of people, making the city-state more liveable for locals, and more attractive for visitors and investors.

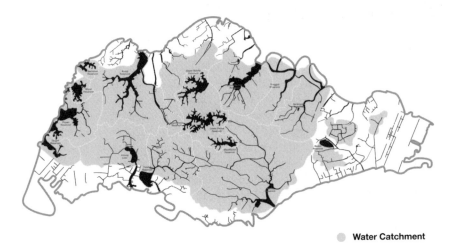

Water Catchment

Figure 8.1 Blue Map of Singapore 2011

In Singapore, overall policy-making, including the one related to water resources, has very much followed a 'think ahead, think again and think across' philosophy (Neo and Chen, 2007).[1] Within this overall coordinated framework, policy- and decision-making processes consider possible important future events as well as their impacts and implications (think ahead), re-evaluate the judgments made in the light of different scenarios and modify them accordingly in order to improve them (think again), and look for experiences and know-how world-wide to enrich the available pool of knowledge (think across). This constitutes a comparative advantage for the country as possible surprises are anticipated and factored in; decisions are reassessed and improved according to global changes, scientific and technological developments and societal attitudes and perceptions; and yielded results and a broad range of potentially relevant scenarios are evaluated. Therefore, apparently serendipitous events are not always unexpected and future changes and challenges are instead determinedly searched for, identified and addressed in advance of their occurrences.

Changes, challenges and strategies on water resources

As the Singapore water story unfolds, it becomes evident that the city-state continues to pursue a development path that would not undermine its scarce land and water resources over the long-term. Following this line of thought, in 2009, the Inter-Ministerial Committee of Sustainable Development identified four strategies to ensure the city-state's continuous and sustained development. These strategies included improving resource use and management efficiencies in energy, water and waste management so that they continue becoming more cost-competitive and efficient; enhancing the physical environment by controlling pollution, increasing green areas and cleaning and beautifying water bodies; engaging and encouraging communities to play their part by adopting more responsible practices, habits and lifestyles; and building technologies and capabilities with the objective to reach sustainable development targets, propel economic growth and export local expertise (MEWR and MND, 2009). The report of this Inter-Ministerial Committee acknowledges that population, commercial and industrial growth are strengths as well as matters of concern because of the impacts they may have on each other under varying conditions. For example, while continuous population growth is expected to have a positive impact on economic growth, water, land and energy resources needed to sustain such growth rates will not increase proportionately, quite the contrary. Population has increased fivefold between 1950 and 2010 (from one to five million) while daily water consumption has increased 12.2 fold (from 32.1 Mgal/day to 380 Mgal/day), putting enormous economic and environmental stress on the system.

Additionally, not only national events but also global and international dynamics will affect the city-state. For instance, the global scarcity of energy and other natural resources is bound to have significant impacts on Singapore as it imports 'almost all [the] resources needs, including basic items such as energy, food and water' (MEWR and MND, 2009: 33) as well as all the resources and raw materials needed for its industries to operate.

Looking for greater natural resource use efficiencies, the Inter-Ministerial Committee has established specific targets to be achieved by 2020 and 2030. For example, energy intensity per dollar GDP is expected to decrease by 20 per cent and 30 per cent from 2005 levels by 2020 and 2030 respectively. Regarding domestic water consumption, the objective is to drive it down from 156 l/capita/day in 2008 to 147 l/capita/day by 2020 and 140 l/capita/day one decade later. To expand water supplies, plans have been made for catchment areas to cover two-thirds of the island's land area (a goal that has already been achieved); develop new fringe catchments; interconnect NEWater plants with the demand areas by pipeline networks; utilize NEWater more intensively and recycle more water; and encourage more industries to replace potable water for non-potable purposes with NEWater or seawater whenever technologically and economically possible.

When developing its plans and strategies, Singapore has recognized the importance of universal principles and paradigms. Nevertheless, it has also acknowledged that these do not automatically lead to improvement unless there is a strong emphasis on context as well as on policy and programme implementation. The city-state has thus traditionally had a very pragmatic view on the management of all its resources. As such, the resulting water policy-making, management, development and governance approaches can provide numerous learning experiences to both developed and developing countries. These series of successful initiatives include, but are in no way limited to, the formulation of strategies reflecting a long-term realistic vision and forward-looking planning exercises; the importance of visionary leadership and strong political will that places pragmatism and general wellbeing over populism; and the effective and efficient coordination and collaboration between ministries, agencies and sectors to reach a common objective. They also highlight the relevance of basing urban development on sound land, water, infrastructure and environmental policies; of putting in place a relevant and stringent legal framework to solve problems stemming from industrial growth and urban sprawl; and of using a wide array of water conservation instruments that consider tariff structures, laws and regulations, reduction and control of unaccounted-for water and public education and awareness raising programmes and campaigns.

Singapore's water story points at the importance of investing in new and innovative technologies that use and reuse all available resources as many times as possible. It also realizes the relevance to put in place policies that encourage the use of such technological advances, including substituting potable water with available alternatives such as marginal quality water, NEWater or desalinated sea water as and when feasible.

There have been many achievements which are unquestionably commendable and exemplary. Nonetheless, with the resource constraints the city-state faces, it will have to run faster to keep up with developments, maintain good rates of economic growth and, above all, respond to societal expectations. While the city-state justifiably prides itself on the enormous progress it has made over the last decades, it is also fully aware of the many challenges it faces, of the old but recurrent issues that need to be improved upon and of the new emerging problems that ought to be addressed in timely manners. As in the past, the motivation to continue striving for excellence and improvement will emerge from domestic social and political expectations as well as from international dynamics. At home, the government will have to respond to the population's increasing hopes and aspirations to lead a better quality of life. As part of a broader community of nations and rapidly changing and increasing competition, Singapore will have to successfully navigate in the uncertain future and constantly changing local and global social, economic, technological and environmental conditions that will be beyond its control.

Unsurprisingly, looking towards the future, and under a rapidly evolving scenario which will be shaped by national and global changes and challenges, management of its limited water resources will become much more complex for the city-state. This will call for the Public Utilities Board (PUB) to continue developing innovative strategies to be able to cover an expected total water demand of 3,460,000 m^3/day (or 0.00346 km^3/day) by 2060, nearly double what it was in 2011 (Puah, 2011). Even for a country like Singapore with an excellent track record of urban water management in recent decades, this would be a complex and most challenging task.

True to its institutional long-term planning tradition, the PUB has already started identifying some key local and global challenges that Singapore and the Board itself are likely to face in the coming years. Among these challenges are climate change, heightened urbanization, rising energy prices and changing expectations by the general public (PUB, personal communication, 2012). These issues are briefly discussed below.

Climate change

The unforeseen and unpredictable impacts of climate change, both temporal and spatial, make it imperative for Singapore to prepare, anticipate, respond and mitigate possible extreme regional weather conditions that may have an impact on water quantity and quality not only in Singapore but also in Johor from where the city-state obtains half of its water supply.

In much of monsoon Asia, where some 80 per cent of annual rainfall occurs in 80–120 hours (not consecutive), current models available are still unable to predict the possible changes in the duration and location of these intense rainfalls. In terms of water resources planning, an important focus has to be on extreme events such as floods and droughts. The disadvantage is that, at the current state of knowledge, it is still not possible to predict with any degree of certainty how the rainfall patterns in Singapore and Johore may change.

The PUB is developing several strategies to ensure that the city-state is as water-resilient as possible, despite potential extreme events, to successfully meet the challenges imposed by the possible impacts of climate change. These include:

- Developing weather-resilient sources (namely desalinated water and NEWater, with the caveat that a serious drought has the potential to directly or indirectly affect all sources of water).
- Continue investing in R&D to further desalination efforts as cost-effectively and as environmentally-friendly as possible.
- Intensifying recycling and reuse practices and use as much NEWater as possible for non-direct potable uses so that potable water is released for domestic uses.
- Planning for long-term infrastructural needs. For example, in anticipation of rising sea levels in the future, the minimum land reclamation levels for newly reclaimed land have been raised by 1 m since late 2011, in addition to the previous level of 1.25 m above the highest recorded tide level observed before 1991.

Increasing urbanization

It is unlikely that Singapore's urbanization trend will reverse; on the contrary, it is actually expected to continue. An expanding metropolis increases the pressure on the PUB to meet the water demands of a larger population for whom drinking water, drainage, wastewater and sanitation services have to be delivered in the most efficient, cost-effective manner and socially acceptable manner. To complicate the current situation even further, the economic demand for water from both the domestic and non-domestic sectors is projected to increase significantly in the future. Restricted by land size, a main concern has to be that the development of infrastructure will be much more difficult and costly as the different agencies will continue competing for scarce land for different purposes.

Rising energy prices

Singapore is both water and energy deficient and thus externally dependent on the two resources. All over the world, the water sector is a major user of energy and vice versa and the city-state is no exception. Energy is needed, for example, for pumping, treating, recycling, desalination and for production of NEWater; and water is needed for energy generation. This close relationship implies that every single strategy and even modest measures to make water and energy use more efficient, will be highly beneficial to the country.

Changing attitudes of the general public

The scope of the water challenge Singapore faces can only be overcome by engaging the population actively in the proposed solutions, chiefly by encouraging it to value water more highly than ever before and thus using it as efficiently as possible. While the PUB has succeeded in communicating to the general population the importance of clean catchments and water conservation, the population still has to fully appreciate that living in a water-deficient place requires major attitudinal changes towards its use as well as more responsible and reduced consumption patterns.

Many efforts have been undertaken to deliver and convey conservation messages that would encourage the population to adopt more water-friendly habits and ideas. For example, before 1997, tariffs (including the water conservation tax) were divided into multiple tiers according to rising consumption levels. Had this price structure been maintained, business costs would have increased disproportionately as tariffs rose, and Singapore's economic competitiveness may have been curtailed. Therefore, the tariff structure was revised on 1 July 1997 with the underlying assumption that water conservation measures should apply equally to all households and businesses (Ng, 1998).

Even now, when water tariffs recover the costs incurred in the production of water, water prices were last revised in the year 2000. This last price increase was to peg the second-tier domestic water price to the marginal cost of new sources of water, which was at that time desalinated water. In our view, the rationale behind this decision is not the best for at least three reasons. First, as management and technology practices have improved over the past 12 years, the marginal cost of desalinated water has come down as well. As such, water prices should be reviewed to account for this reduction in treatment costs, while also catering for other costs incurred in the production and supply of water in Singapore, e.g. distribution network for water, and used water collection and treatment costs.

Second, water tariffs have not changed since 2000, but electricity tariffs have increased significantly. Thus, if only inflation is concerned, and assuming a household is using a certain amount of water, its water bill in real terms has declined by over 25 per cent because of inflation during the intervening period. In addition, the average monthly household income in 2000 was $4,988, increasing to $7,214 by 2012, which means that the water bill as a percentage of household income has steadily declined since 2000.[2] For most households, it is likely that the bill now represents less than 1 per cent of their income, and is thus a minor item of expenditure. It is also interesting to note that Singapore's total bottled water sales have doubled from US$65 million in 2000 to US$131 million in 2012. Thus, even though the PUB provides excellent high quality tap water to every home, total bottled water sales are exploding even when bottled water costs hundreds of times more and has no discernible health benefit.

Third, the low cost of water is reflected by Singapore's per capita water consumption which was around 153 l/capita/day in 2011. Both the current consumption rate and the target for 2030 (140 l/capita/day) could be further reduced, especially as the city-state is water deficient. These figures do not compare favourably with many European cities where the consumption rates are close to 100 l/capita/day and which are likely to break this barrier by 2030. Much of these reductions have occurred due to pricing, which has radically changed the water use patterns of the population, with the added benefit that no adverse health or social impacts have been reported because of reduced water consumption.

Given its limited natural endowments, water security considerations and future uncertainties emerging from global competition for resources and also potential impacts of climate change, further policy measures are needed to significantly reduce per capita consumption and industrial water demands. Since economic instruments have proven successful in reshaping consumption patterns and human behaviour, such measures should receive priority consideration to further bring down water consumption.

Innovative technologies supported by policy and management innovations

In order to meet the expected future water demands from the domestic and non-domestic sectors, non-conventional sources are likely to acquire an increasingly relevant role. By 2060, the production of NEWater is expected to triple to meet 50 per cent of all demand, compared to 30 per cent in 2011. Regarding desalination, the PUB's objective is to intensify capacity by ten-fold to meet 30 per cent of the long-term water needs mostly coming from commercial and industrial users. They are estimated to increase their consumption share from 40 per cent in 2011 to 80 per cent in 2060. This makes the need for innovative, cost-effective and efficient technology as well as investments on R&D to make processes more efficient and less energy intensive all the more essential (Loh, 2011a).

Already a successful and inventive policy, the production of NEWater has proven to be of utmost importance as it covers the water demands of the development-propelling commercial and industrial sectors. Even more importantly, it offers the possibility to achieve self-sufficiency in the near future.

In 2011, the existing plants could produce about 117 Mgal/day of NEWater: Bedok (18 Mgal/day), Kranji (17 Mgal/day), Ulu Pandan (32 Mgal/day) and Changi (50 Mgal/day). The NEWater produced was commercially and industrially utilized in different production processes where some 13.5 Mgal/day were used for wafer fabrication; 19.8 Mgal/day for manufacturing industries including petrochemicals, chemical and electronics; 4.3 Mgal/day for commercial buildings; and 3.2 Mgal/day for other purposes (Loh, 2011b). NEWater is also used for cooling towers, for general washing and toilet flushing in commercial buildings, etc.

The PUB's target is to cover an expected demand of 760 Mgal/day by 2060. In order to achieve this objective, the agency has been working on a series of strategies to increase the amount of water available for recycling and to expand its use more significantly for non-potable uses, releasing more drinking water for potable uses. These measures include the use of novel technologies; infrastructural improvements and developments; pursuing greater recovery rates; improving industrial efficiency in recycling even larger volumes of water and, whenever possible, make industrial premises water neutral. This later point includes reducing evaporative water losses so that more water is available for recycling, especially for industries on Jurong Island (PUB, personal communication, 2012).

Despite noteworthy achievements, associated challenges are manifold and relate not only to the economics of efficient R&D but also to innovative, cost-effective and efficient technology. They also relate to social and environmental aspects, which are of equal importance. Accordingly, in social terms, the PUB has been working for years on education and information programmes to engage the population as part of several initiatives, mostly on water conservation, with a main focus on public perception and acceptance of the direct and indirect potable uses of NEWater.

Regarding environmental issues, activities have focused on reducing dependency on natural resources and on adopting economic strategies that factor in the importance of supplying clean water at affordable prices (PUB, personal communication, 2012). In addition, the expansion of desalination capacity and related environmental concerns such as energy use and the disposal of brine in an environmentally acceptable way, are also being addressed. In 2011, the energy requirements to produce desalinated water using reverse osmosis (RO) remained high at approximately 3.5 kWh/m^3. According to the PUB, the journey to low energy seawater desalination would include moving from the current 3.5 kWh/m^3 consumption to 1.5 kWh/m^3 in the short-term, for example, through variable salinity processes (1.7 kWh/m^3), membrane distillation (with waste heat) (1.0 kWh/m^3) and electrochemical desalting (1.5 kWh/m^3). In the long-term, the objective is to reduce energy requirements to 0.75 kWh/m^3, mostly through technological (Puah, 2011) and management advancements.

There has been a significant investment in R&D to achieve breakthroughs in all types of technology. Regarding desalination, studies to make desalination processes more energy efficient include, *inter alia*, electrochemical desalting to reduce the energy consumption to less than half of what current membrane-based desalination methods use as well as efforts to maximize membrane strength to help them resist the high-pressure desalination environment. According to researchers working on this latter initiative, laboratory-developed membranes are approximately 40 per cent more permeable compared to more commercial brackish water RO membranes, which would make them considerably less energy intensive. For saltier seawater, the performance gap is reported to be about one order of magnitude more permeable than typical seawater RO membranes. Even though the main limitation is to scale them up in order to make them attractive at industrial levels, it is estimated that it will be possible to use the membranes in existing desalination facilities, with the resulting reduction in terms of costs and energy, within only one to two years. Other studies focus on biomimicry of natural desalination systems such as mangrove plants and marine fishes (National Climate Change Secretariat, 2012; PUB, 2012).

Through the Environment and Water Industry Programme Office (EWI), the PUB leads all R&D efforts. This is an interagency body that also includes the Economic Development Board (EDB), International Enterprise Singapore (IES), the enterprise development agency SPRING Singapore as well as the National University of Singapore, Nanyang Technological University and the Agency for Science, Technology and Research (A*STAR). Consistent with the cross-sectoral and multidisciplinary approach adopted to give response to the city-state's challenges, government strategies carried out through the EWI Programme Office integrates policy and implementation frameworks that stretch across the various agencies involved in the development of the water industry (PUB, 2012).

In addition to the above activities, investments in overall water research would benefit tremendously if data and information could be made readily available to universities and research institutions. This would encourage research on broader topics related to water planning, policy-making and governance as well as specific ones such as use of economic instruments, social and environmental issues and social perceptions and attitudes.

There is an excellent example from Brazil when, a decade ago, the National Water Agency (ANA) made water-related data available freely on the Internet. This contributed to the renaissance of water research in the Brazilian universities and research institutions which ultimately helped to improve water management practices in the country. Access to water-related data on the Internet in Singapore would unleash a wave of research activities in different areas that would complement the research conducted by and for the PUB, attracting scholars from various disciplines and with different mindsets. This would add significantly to the generation, synthesis, application and dissemination of knowledge in the overall field of water resources in the city-state.

Further thoughts

Looking back, Singapore's main challenge has been the limited water supplies available to cover the actual and estimated growing domestic, commercial and industrial demands at different times in history. By facing this challenge, the city-state has transformed the potential crisis into an opportunity that has led to the development, implementation, fine-tuning and improvement of multiple supply and demand management strategies. What is truly commendable is that in spite of its scarce water and land resources, Singapore has not limited the growth of its urban, commercial and industrial sectors. On the contrary, it has worked with determination to produce sufficient water, even anticipating and planning for the future needs by relentlessly searching for new alternatives and opportunities. The city-state has invested, and continues to invest, very heavily on research and development to build up the necessary know-how and technology on water production. These investments, together with pragmatic management practices, have propelled Singapore towards a solid path to attain overall development, social and economic growth, better quality of life for its population and protection of the environment.

Nonetheless, no production of water would ever be enough unless managed and used wisely, a main issue that calls for serious reflection in the implementation of stricter water conservation measures. No matter how much water is drawn from the four taps, the amount will never be sufficient if not consumed more efficiently.

Finally, during its journey for water self-sufficiency, Singapore has gained in terms of resilience, determination, inventiveness, innovation and the engrained aspiration to look beyond the obvious for better alternatives. All of it has been possible because of strong and consistent political support from the highest levels and an excellent cadre of civil servants.

As noted earlier in this book, while everything can be improved, so far, Singapore has developed an excellent laboratory for public policies of water resources as well as their implementation. All too aware of the challenges posed by heightened urban sprawl and the limited availability of natural endowments, the island's political leadership has already started to think ahead, think again and think across to address the competition for scarce resources at the global level and the threat of environmental degradation. Considering the city-state's institutional legacy, structures and past performance, one can only expect that innovative strategies, policies and management alternatives will be continually developed and then implemented. Nevertheless, only time will tell to what extent the Singapore water story will be appraised and capitalized on and what will be the path the city-state and its people decide to take to continue to strive for increasingly better human and natural environments.

Notes

1 Setting the foundations

1 The PAP has been the ruling party in Singapore since 1959, when it first came to power.
2 Hawkers refer to food sellers.
3 Spelling of Johore as it appears in the agreement.

2 Water and urban development

1 The five-year plan is a compilation of policy statements issued by members of its Central Executive Committee before the 1959 elections.
2 Singapore and Malaysia had a common currency (the Malaya dollar) until 12 June 1967, when both the governments decided to issue their own currencies (The Straits Times, 18 August 1966: 1; 11 December 1966: 1).
3 According to the Singapore Department of Statistics, the total population in 1971 was 2.112 million people (Singapore Department of Statistics, http://www.singstat.gov.sg/stats/themes/people/hist/popn.html, accessed on 14 September 2012).
4 See URA, http://www.ura.gov.sg/about/ura-history.htm (accessed on 29 August 2012).
5 Singapore Statutes are available at http://statutes.agc.gov.sg/ (accessed on 15 September 2012).
6 See also MEWR, http://app.mewr.gov.sg/web/Contents/Contents.aspx?ContId=1342 (accessed on 14 September 2012).

3 Regulatory instruments and institutions for water pollution control

1 See NEA, http://www.nea.gov.sg/cms/pcd/EPDAnnualReport.pdf (accessed on 7 September 2012).
2 See URA, http://www.ura.gov.sg/land_use_planning (accessed on 14 September 2012).
3 See NEA, http://www.nea.gov.sg/cms/pcd/EPDAnnualReport.pdf (accessed on 7 September 2012).
4 See also http://app.mewr.gov.sg/web/Contents/Contents.aspx?ContId=1342 (accessed on 14 September 2012).
5 See PUB, http://www.pub.gov.sg/general/code/Pages/default.aspx (accessed on 14 September 2012).
6 See http://statutes.agc.gov.sg/aol/search/display/view.w3p;page=0;query=DocId%3A138d92b9-9a21-4649-b14d-97552a8af9a1%20%20Status%3Ainforce%20Depth%3A0;rec=0#legis (accessed on 7 September 2012).
7 See NEA, http://app2.nea.gov.sg/legislation.aspx (accessed on 6 September 2012).
8 See http://statutes.agc.gov.sg/aol/search/display/view.w3p;page=0;query=DocId%3A8

615ccd4-a019-485d-aa9e-d858e4e246c5%20Depth%3A0;rec=0;resUrl=http%3A%2
F%2Fstatutes.agc.gov.sg%2Faol%2Fbrowse%2FtitleResults.w3p%3Bletter%3DE%3B
type%3DactsAll;whole=yes (accessed on 14 September 2012).

9 See http://statutes.agc.gov.sg/aol/search/display/view.w3p;page=0;query=CompId%3
Aadc249d9-fcff-4dd0-9f3b-4be8d8d22ca2%20ValidTime%3A20120524000000%20
TransactionTime%3A20120524000000;rec=0 (accessed on 14 September 2012).

10 See NEA, http://app2.nea.gov.sg/semakaulandfill.aspx (accessed on 7 September 2012).

4 Managing water demands

1 Confrontation or Konfrontasi was a conflict between Indonesia and Malaysia that took
place mainly on the island of Borneo. For more information, see Lee (1998).

2 In an Appraisal of the Singapore Sewerage Project by the Ministry of Overseas
Development and Overseas Development Administration, it is mentioned that the
sewerage extension and improvement programme would be for the period 1968 through
June 1972. It would include the construction of sewers, pumping stations in nine areas,
modifications and expansion of three existing treatment plants and miscellaneous
improvements and extensions to existing pumping stations and sewers. The total cost
of the proposed project is mention to have been estimated at $67.2 million (equivalent
to US$22.4 million) including interest during construction and that the proposed loan
for US$6 million would be used to meet about 77 per cent of the foreign exchange costs.
The balance of the project expenditures would be covered by appropriations from the
government's development funds. The loan would be for a period of 20 years including five
years' grace. Sewerage was a minor part of the second five-year development plan for the
1966–1970 period in terms of money (3.6 per cent of public capital expenditure) but it was
of fundamental importance for the industrial estates and the housing and urban renewal
programmes. Up to that moment, the housing and industrial estate programmes had been
considered as 'outstanding successes' and there was nothing that would indicate that they
would fall short of their goals (National Archives of The United Kingdom, n.d.).

3 The sanitary appliance fee and the waterborne fee are levied to offset the cost of
treating used water and for operating and maintaining the wastewater network. The
sanitary appliance fee is a fixed component based on the number of sanitary fittings in
each premise whereas the waterborne fee is charged based on the volume of water used
in any given premise (for more information, see PUB, http://www.pub.gov.sg/general/
Pages/WaterTariff.aspx, accessed on 31 August 2012).

4 Thimbles are flat metal discs with a hole in the middle and are placed in pipes to reduce
the flow-rate and thereby reduce water use.

5 See also http://www.ies.org.sg/e-newsletter/PUBMWELS.pdf.

6 See Attorney-General's Chambers, http://statutes.agc.gov.sg/aol/search/display/view.
w3p;page=0;query=DocId%3A138d92b9-9a21-4649-b14d-97552a8af9a1%20%20Sta
tus%3Ainforce%20Depth%3A0;rec=0 (accessed on 8 September 2012).

5 Education and information strategies for water conservation

1 See Singapore statistics, http://www.singstat.gov.sg/pubn/popn/c2010acr.pdf (accessed
on 24 May 2012).

2 See Singapore statistics, http://www.singstat.gov.sg/stats/themes/economy/hist/gdp.
html (accessed on 14 September 2012).

3 See Singapore statistics, www.singstat.gov.sg/stats/keyind.html#keyind (accessed on
24 May 2012).

4 See Singapore statistics, www.singstat.gov.sg/stats/keyind.html#keyind (accessed on
24 May 2012).

5 See http://earthtrends.wri.org/pdf_library/country_profiles/pop_cou_702.pdf (accessed
on 24 May 2012).

238   *Notes*

6 See Singapore statistics, http://www.singstat.gov.sg/stats/themes/people/popnindicators.
    pdf (accessed on 14 September 2012).
7 See Singapore statistics, www.singstat.gov.sg/stats/keyind.html#keyind (accessed on
    24 May 2012).
8 The merits of the Speakers' Corner is beyond the scope of this paper and thus will not
    be discussed.
9 See PUB, http://www.pub.gov.sg/LongTermWaterPlans/cve_love.html (accessed on
    14 September 2012).
10 See Ministry of the Environment and Water Resources, http://www.mewr.gov.sg/
    sgp2012/about.htm (accessed on 4 September 2012).
11 http://migration.ucdavis.edu/mn/more.php?id=929_0_3_0 (accessed on 4 September
    2012).
12 For more information on NEWater, see PUB, http://www.pub.gov.sg/ABOUT/
    HISTORYFUTURE/Pages/NEWater.aspx (accessed on 4 September 2012).
13 See PUB, http://www.pub.gov.sg/conserve/Households/Pages/default.aspx (accessed
    on 15 September 2012).
14 See PUB, http://www.pub.gov.sg/conserve/CommercialOperatorsAndOther/Pages/
    default.aspx (accessed on 15 September 2012).

6 Cleaning of the Singapore River and Kallang Basin

1 The cost estimates presented for the cleaning of the Singapore River through the years
    do not take inflation into consideration.

7 Views of the media on the Singapore–Malaysia water relationship

1 An earlier version of this chapter was written in collaboration with Kimberly Pobre.
2 The older spelling, 'Johore', is used in some references.
3 http://statutes.agc.gov.sg/aol/search/display/view.w3p;query=CapAct%3A338%20Ty
    pe%3Auact,areved;rec=0;resUrl=http%3A%2F%2Fstatutes.agc.gov.sg%2Faol%2Fse
    arch%2Fsummary%2Fresults.w3p%3Bquery%3DCapAct%253A338%2520Type%25
    3Auact,areved;whole=yes (accessed on 1 September 2012).
4 The chronology of the procedure and submissions of the Parties regarding the
    sovereignty over Pedra Branca/Pulau Batu Puteh, Middle Rocks and South Ledge
    (Malaysia/Singapore) can be found in the Summary of the Judgment of 23 May 2008,
    International Court of Justice, 23 May 2008 (ICJ, 2008).
5 See Ministry of Foreign Affairs, http://www.mfa.gov.sg/internet/press/pedra/faq.html
    (accessed on 10 September 2010).

8 Looking ahead

1 Neo and Chen (2007) define these three concepts as follows: 'Think ahead is the capability
    to identify future developments in the environment, understand their implications on
    important socio-economic goals, and identify the strategic investments and options
    required to enable a society to exploit new opportunities and deal with potential threats'
    (p. 30). 'Think again is the capability to confront the current realities regarding the
    performance of existing strategies, policies and programs, and then to redesign them to
    achieve better quality and results' (pp. 35–36). 'Think across is the capability to cross
    traditional borders and boundaries in order to learn from the experience of others so that
    good ideas may be adopted and customized to enable new and innovative policies or
    programs to be experimented with and institutionalized' (p. 40).
2 See Singapore statistics, key indicators of residence households, http://www.singstat.
    gov.sg/stats/themes/people/ hhldincome.html (accessed on 27 September 2012).

Bibliography

Abdulla, S.A. (2003a) 'Booklet will tell the truth', *New Straits Times*, 20 July.

Abdulla, S.A. (2003b) 'Singapore's reluctance to pay fair price for water puzzling, says booklet', *New Straits Times*, 21 July.

Abdullah, A. (1998a) 'Understanding over water supply to Singapore after 2061 reached', *New Straits Times*, 18 April.

Abdullah, A. (1998b) 'Republic's RM 15.2b loan for water not needed', *New Straits Times*, 18 December.

Abdullah, A. (1998c) 'Singapore exploits neighbours' weaknesses, says PM', *New Straits Times*, 15 September.

Abdullah, A. and Bingkasan, J. (1998) 'KL and Singapore set to resolve water supply issues', *New Straits Times*, 4 April.

Abdullah, A. and Rajendram, S. (1998) 'No agreement yet on water deal', *New Straits Times*, 8 July.

Abdullah, F. (2002a) 'Water issue: Malaysia to stop talks with Singapore', *New Straits Times*, 30 November.

Abdullah, F. (2002b) 'Singapore should be sincere', *News Straits Times*, 19 November.

Abdullah, F. (2003a) 'Bridge to replace causeway', *New Straits Times*, 2 August.

Abdullah, F. (2003b) 'Malaysia to stop supplying raw water to Singapore in 2061, says Syed Hamid', *New Straits Times*, 29 January.

Abdullah, F. and Megan M.K. (2002) 'More favour than trade', *New Straits Times*, 7 August.

Abdullah, Z. (2003) 'Singapore will try to delay the settlement of water issues', *Utusan Malaysia*, 4 August (translated from Malay into English).

Abu Bakar, Z. and Chow, K.H. (2002) 'PM: Singapore changing stand', *New Straits Times*, 2 November.

Agreement as to Certain Water Rights in Johore between the Sultan of Johore and the Municipal Commissioners of the Town of Singapore, signed in Johore in 5 December 1927. Singapore Parliamentary Debates, Official Report, vol. 75, cols. 2585–2604, Annex A, 25 January 2003.

Agreement between the Government of the State of Johor and the Public Utilities Board of the Republic of Singapore, signed in Johore on 24 November 1990. Singapore Parliamentary Debates, Official Report, vol. 75, cols. 2665–2730, Annex D, 25 January 2003.

Ahmad, A.R. and Latiff, R.A. (2002) 'Malaysia seeks assurance', *New Straits Times*, 9 March.

Ahmad, E. (2002) 'Malaysian media takes potshots at NEWater', *The Straits Times*, 6 September.

Ahmad, R. (2002a) 'Malaysia reveals asking price for water; 60 sen per 1000 gallons till 2007, after which it will be raised to RM 3, says Foreign Minister. S'pore pays 3 sen now', *The Straits Times*, 8 July.

Ahmad, R. (2002b) 'KL ultimatum gives S'pore "one last chance" on water', *The Straits Times*, 14 October.

Ahmad, R. (2002c) 'KL turns heat on Singapore ahead of JB talks', *The Straits Times*, 15 October.

Ahmad, R. (2002d) 'KL mulls over new law to scrap water accords', *The Straits Times*, 25 October.

Ahmad, R. (2002e) 'KL hints at other options to settle water issue', *The Straits Times*, 3 December.

Ahmad, R. (2002f) 'KL will not go to court over water: Syed Hamid', *The Straits Times*, 31 December.

Ahmad, R. (2003a) 'Changes in control of water resources won't affect Singapore deals', *The Straits Times*, 13 June.

Ahmad, R. (2003b) 'Water supply deal will remain: Mahatir', *The Straits Times*, 2 August.

Ahmad, R. and Latiff, R.A. (2002) 'Malaysia seeks assurance', *New Straits Times*, 9 March.

Alias, M. and Said, H. (2003) 'PM: expedite arbitration', *Berita Harian*, 4 August (translated from Malay into English).

Andina-Díaz, A. (2007) 'Reinforcement *vs* change: the political influence of the media', *Public Choice*, vol. 31, no. 1–2, pp. 65–81.

Ang, P.H. (2002) 'The media and the flow of information', in D. Da Cunha (ed.) *Singapore in the New Millennium: Challenges Facing the City State*, Institute of Southeast Asian Studies, Singapore, pp. 243–268.

Ang, P.H. (2007) *Singapore Media*. Available at: http://journalism.sg/wp-content/uploads/2007/09/ang-penghwa-2007-singapore-media.pdf (accessed on 10 January 2011).

Ang, W.M. and Lee, H.S. (1995) 'Govt concerned but won't cap manufacturers' water usage', *The Business Times*, 20 October.

Anti-Pollution Unit (APU) (1979) *Annual Report 1978*, Government Printing Office, Singapore.

Anti-Pollution Unit (APU) (1981) *Annual Report 1980*, Government Printing Office, Singapore.

Ariff, B.B.M. (1959) *Parliament Debate on the Suspension and Transfer of Functions Bill for the City Council*, Parliament no. 0, Session no. 1, vol. 11, Sitting no. 2, Sitting date: 15 July, Hansard, Singapore.

Arof, F. (2002) 'Rais: Malaysia has the right to terminate the water agreement', *Utusan Malaysia*, 25 October (translated from Malay into English).

Atan, H. (2003) 'Malaysia wants arbitration soon', *New Straits Times*, 4 August.

Aziz, A. and Haron, S. (1998) 'Pahang to consult NWC on sale of water to Singapore', *New Straits Times*, 25 June.

Aziz, M. (2003) 'We might as well give water free', *New Straits Times*, 24 July.

Bakar, W.H. (2003) 'Singapore urged to stop blaming others', *Berita Harian*, 21 July.

Balakrishnan, V. (2012) *Opening Speech*, Singapore International Water Week 2012, Singapore.

Barker, E.W. (1971) *Debate on the President's Address*, Parliament no. 2, Session no. 2, vol. 31, Sitting no. 6, Sitting date: 5 August, Hansard, Singapore.

Barker, E.W. (1976) *Budget, Ministry of the Environment*, Parliament no. 3, Session no. 2, vol. 35, Sitting no. 7, Sitting date: 22 March, Hansard, Singapore.

BBC (13 July 1998) 'Conditions attached to water supply to Singapore'.

BBC (16 September 1998) 'Premier Mahatir criticizes Singapore for raising "old issues"'.

BBC (7 June 1999) 'Malaysian Premier discusses package of issues between Malaysia and Singapore'.

BBC (9 June 1999) 'Premier discusses package of issues between Malaysia, Singapore'.

BBC (22 January 2002) 'Malaysia, Singapore unable to agree on water supply price'.

BBC (28 January 2002) 'Singapore denies Malaysian allegations water agreement favour republic'.

BBC (12 March 2002) 'Singapore studies new Malaysian proposals in water price row'.

BBC (1 July 2002) 'Singapore, Malaysian foreign ministers set for water, economic talks'.

BBC (3 July 2002) 'Malaysia, Singapore agree to treat water issue separately'.

BBC (5 July 2002) 'Malaysian daily views negotiations with Singapore on water supply'.

BBC (1 November 2002) 'Singapore blames Malaysia for breakdown on water talks'.

BBC (16 July 2003) 'Malaysian deputy premier rejects Singapore claims on water adverts'.

BBC (21 July 2003) 'Malaysia issues booklet on water dispute with Singapore'.

BBC (22 July 2003) 'Malaysian FM on Singapore water dispute, Burma, North Korea'.

BBC (29 July 2003) 'Singapore places advertisements on water dispute in five Malaysian newspapers'.

BBC (14 December 2004) 'Malaysia, Singapore to resume talks on unresolved issues'.

BBC (12 February 2007) 'Malaysia Mahatir urges renegotiations with Singapore over water supply issue'.

Begum, F. (2002) 'Johor water may be sold to other states', *The Star*, 11 October.

Berita Harian (6 July 2002) 'Malaysia will not compromise the price of water' (translated from Malay into English).

Berita Harian (9 July 2002) 'Review of water prices to be discussed in August' (translated from Malay into English).

Berita Harian (7 August 2002) 'No assurance that water issue will be resolved' (translated from Malay into English).

Berita Harian (5 September 2002) 'Singapore ego prevents smooth negotiations' (translated from Malay into English).

Berita Harian (19 September 2002) 'P.C.A. is ready to arbitrate water pricing problem' (translated from Malay into English).

Berita Harian (19 October 2002) 'Singapore is responsible for water talks deadlock' (translated from Malay into English).

Berita Harian (25 October 2002a) 'Malaysia can develop a new act to solve the water problems' (translated from Malay into English).

Berita Harian (25 October 2002b) 'Malaysia should increase the price of water' (translated from Malay into English).

Berita Harian (26 October 2002) 'Abim wants the water issue resolved soon' (translated from Malay into English).

Berita Harian (2 November 2002) 'Singapore needs to be serious to resolve water issues' (translated from Malay into English).

Berita Harian (4 November 2002) 'Water issues: people want KL to support Johor' (translated from Malay into English).

Berita Harian (6 November 2002) 'Johor will take control of water treatment plants' (translated from Malay into English).

Berita Harian (18 November 2002) 'Water: Singapore gives two conditions to continue negotiations' (translated from Malay into English).

Berita Harian (22 November 2002a) 'Water: Singapore deliberately delaying negotiations' (translated from Malay into English).

Berita Harian (22 November 2002b) 'Malaysia's stand is considered the cause of the deadlock' (translated from Malay into English).

Berita Harian (28 January 2003) 'Water issues: Singapore should stop pointing at others' (translated from Malay into English).

Berita Harian (30 January 2003) 'Malaysia has the right to stop water supply from 2061' (translated from Malay into English).

Berita Harian (1 February 2003) 'Sovereignty is the issue for Singapore' (translated from Malay into English).

Berita Harian (10 February 2003) 'Singapore confident of its own production of water' (translated from Malay into English).

Berita Harian (1 March 2003) 'Singapore wants to continue relations with Malaysia' (translated from Malay into English).

Berita Harian (17 March 2003) 'Singapore publishes book on water issues' (translated from Malay into English).

Berita Harian (21 July 2003) 'Harmony with the water issue' (translated from Malay into English).

Berita Harian (26 July 2003) 'Water advertisement: Singapore continues to broadcast the wrong facts' (translated from Malay into English).

Berita Harian (28 July 2003) 'MTEN broadcasts in response to Singapore's misreports in AWSJ' (translated from Malay into English).

Berita Harian (29 July 2003) 'Mandarin version of the water book will start selling tomorrow' (translated from Malay into English).

Berita Minggun (3 November 2002a) 'Solving the water issue would be more complicated if taken to court' (translated from Malay into English).

Berita Minggun (3 November 2002b) 'Water issue to be discussed in Parliament next week' (translated from Malay into English).

Bernama (6 August 2002) 'S'pore must be reasonable over price of water, says Dr Mahathir'.

Bernama (11 October 2002) 'Malaysia to backdate water dues'.

Binnie & Partners (1981) *Report on Singapore Water Supply*, National University of Singapore, Singapore.

Birmingham Post (4 August 1998) 'Malaysia threatens to cut Singapore's water supply'.

Boey, D. (1998) 'S'pore still keen to buy M'sian water: Yock Suan', *The Business Times*, 4 August.

Boey, D. (2003a) 'Hyflux in deal to draw water from air', *The Business Times*, 7 January.

Boey, D. (2003b) 'Govt to keep water tariffs affordable: Swee Say', *The Business Times*, 20 March.

Buenas, D. (2003) 'S'pore a listing hub for water firms?', *The Business Times*, 24 June.

Business Times, The (4 November 1981) 'Don't use water like water'.

Business Times, The (14 March 1995) 'Singapore's water charges to rise sharply in coming years'.

Business Times, The (4 December 1996) 'Market drifts in the absence of overseas leads'.

Business Times, The (4 June 1997) 'Invest in water desalination plants says MP'.

Business Times, The (6 June 1997) 'Framework for wider cooperation'.

Business Times, The (30 June 1998) 'S'pore, KL have not reached accord on water'.

Business Times, The (5 August 1998) 'US Filter, a water treatment firm, opens HQ here'.

Business Times, The (18 December 1998) 'KL passes up US$4B S'pore loan, ties water to other issues KL seen sorting out Clob transfer soon'.

Business Times, The (11 June 1999) 'Looking beyond M'sia for water'.

Business Times, The (20 January 2001) 'Hyflux opens $4m R&D centre, plant at Changi'.

Business Times, The (5 September 2001) 'SM Lee, Dr M strike in-principle accord on bilateral issues'.

Business Times, The (1 February 2002) 'Malaysia will insist on higher rate for water to S'pore: Dr M; A review of the in-principle pact reached last September now looks likely'.

Business Times, The (4 March 2002) 'Johor to give KL report on S'pore reclamation'.

Business Times, The (12 April 2002) 'A bilateral reality gap'.

Business Times, The (17 June 2002) 'PM expects progress in resolving bilateral issues with KL'.

Business Times, The (26 July 2002) 'Good enough to quench the thirst'.

Business Times, The (14 October 2002) 'KL may use legal means over water issue'.

Business Times, The (24 October 2002) 'PM: Singapore reluctant to solve water issues'.

Business Times, The (1 November 2002) 'Johor confirmed 3 sen per price with Linggiu dam accord'.

Business Times, The (2 November 2002) 'Water dispute may have to be resolved in court: Syed Hamid'.

Business Times, The (21 November 2002a) 'Little is expected of next round of water talks'.

Business Times, The (21 November 2002b) 'UE unit clinches $200m PUB contract'.

Business Times, The (7 January 2003) 'Johor to stop buying S'pore treated water'.

Business Times, The (6 February 2003) 'The price of sovereignty'.

Business Times, The (15 February 2003) 'Be my Valentine, S'pore: Dr M'.

Business Times, The (6 December 2003) 'Sewage attraction'.

Business Times, The (7 February 2004) 'There's a new spirit in Malaysia now'.

Business Times, The (6 June 2005) 'Water conference in July'.

Business Times, The (7 December 2005b) 'PUB joins exclusive water research group'.

Business Times, The (21 December 2005) 'United Engineers bags PUB water treatment job'.

Business World (7 August 2009) 'Singapore national day'.

Calderon, M.E. (2008) 'Investment opportunities offered by Singapore', *Business World*, 27 July.

Caponera, D.A. (2003) *National and International Water Law and Administration*, Kluwer Law International, The Hague.

Chan, H.C. (1976) 'The role of parliamentary politicians in Singapore', *Legislative Studies Quarterly*, vol. 1, no. 3, pp. 423–441.

Chan, H.C. (1991) 'Political developments, 1965–1979', in E.C.T. Chew and E. Lee (eds) *A History of Singapore*, Oxford University Press, Singapore, pp. 157–181.

Chan, Y.K. (2001) 'PUB tapping all possible water sources', *The Straits Times*, 18 September.

Chang, C.Y., Ng, B.Y. and Singh, P. (2005) *Roundtable on Singapore–Malaysia Relations: Mending Fences and Making Good Neighbors*, Institute of Southeast Asian Studies, Singapore.

Cheah, C.S. (2002a) 'Malaysia just as eager to solve bilateral issues', *New Straits Times*, 29 January.

Cheah, C.S. (2002b) 'Send protest note if reclamation work affects deep water line', *New Straits Times*, 12 March.

Chen, A.Y. (1991) 'The mass media, 1819–1980', in E.C.T. Chew and E. Lee (eds) *A History of Singapore*, Oxford University Press, Singapore, pp. 288–311.

Cheong, K.H. (2008) *Achieving Sustainable Urban Development*, Ethos, Civil Services College, Singapore

Cheong-Chua, K.H. (1995) 'Urban land use planning in Singapore: towards a tropical city of excellence', in G.L. Ooi (ed.) *Environment and the City: Sharing Singapore's Experience and Future Challenges*, Times Academic Press for Institute of Policy Studies, Singapore, pp. 102–128.

Chew, C.S (1973) 'Key elements in the urban renewal of Singapore', in C.P. Chye (ed.) *Planning in Singapore, Selected Aspects and Issues*, Chopmen Enterprises, Singapore, pp. 32–44.

Chew, C.T. and Lee, E. (1991) *A History of Singapore*, Oxford University Press, Singapore.

Chia, I. (2003) 'Explore peaceful ways to resolve water dispute amicably', *The Star*, 11 February.

Chia, L.S. and Chionh, Y.H. (1987) 'Singapore', in L.S. Chia (ed.) *Environmental Management in Southeast Asia*, NUS Press, Singapore, pp. 109–168.

Chia, L.S. and Chionh, Y.H. (1987) *Environmental Management in Southeast Asia*, NUS Press, Singapore.

Chia, P. (1971a) 'Question is: when will the cuts begin', *The Straits Times*, 8 May.

Chia, P. (1971b) '"Don't waste water" plea gets good public response', *The Straits Times*, 13 May.

Chiang, K.M. (1986) *The Cleaning up of Singapore River and the Kallang Basin Catchments*, COBSEA Workshop on Cleaning-up of Urban River, Ministry of the Environment of Singapore and UNEP, 14–16 January, Singapore.

Chin, C.C. (1999) 'It's up to Singapore to offer solutions to outstanding problems', *New Straits Times*, 23 February.

Chin, K.L. (2002) 'Rais: study on law to protect country's water, natural resources', *New Straits Times*, 3 November.

Ching, L. (2010) 'Eliminating yuck: a simple exposition of media and social change in water reuse policies', *International Journal of Water Resources Development*, vol. 26, no. 1, pp. 111–124.

Choay, F. and Merlin, P. (eds) (2005) *Dictionnaire de l'urbanisme et de l'aménagement*, Quadrige, Press Universitaires de France, Paris, p. 174. Quoted in X. Guillot (2008) 'Vertical living and the garden city', in T.C. Wong, B. Yuen and C. Goldblum (eds) *Spatial Planning for a Sustainable Singapore*, Springer, Singapore, p. 153.

Chong, C.S. (2000) 'Johor's own water by 2003', *New Straits Times*, 22 December.

Chong, C.S. (2002) 'MB: Singapore's stand on water issue a stumbling block to ties', *New Straits Times*, 8 September.

Chong, T. (2005) *Civil Society in Singapore: Reviewing Concepts in the Literature*, Institute of Southeast Asian Studies, Working Paper, Singapore.

Choong, M.Y. (2006) 'S'pore No. 1 in managing water resources: expert', *The Straits Times*, 23 August.

Chou, L.M. (1998) 'The cleaning of Singapore River and the Kallang Basin: approaches, methods, investments and benefits', *Ocean and Coastal Management*, vol. 38, no. 2, pp. 133–145.

Chua, J. (2005a) 'S'pore to increase water catchment area', *The Business Times*, 14 January.

Chua, J. (2005b) 'Govt to pour $1.5b into water-related projects', *The Business Times*, 14 September.

Chua, L.H. (2002a) 'S'pore denies unfair deals claims', *The Straits Times*, 28 January.

Chua, L.H. (2002b) 'New bilateral talks begin in KL', *The Straits Times*, 1 July.

Chua, L.H. (2002c) 'KL seeking to settle water pricing separately', *The Straits Times*, 3 July.

Chua, L.H. (2002d) 'Interdependence at the heart of water issue', *The Straits Times*, 4 September.

Chua, L.H. (2002e) 'S'pore to KL: remember your promises; Malaysia repeatedly vowed to supply water', *The Straits Times*, 26 October.

Chua, S.C. (1973) *Singapore Success Story, Speech at the First Australian Seminar on Litter Pollution at Perth on 6 October 1972, Western Australia*, Ministry of the Environment, Singapore.

Chuang, P.M. (2002a) 'NEWater gets panel's clearance', *The Business Times*, 12 July.

Chuang, P.M. (2002b) 'NEWater purer than PUB's', *The Business Times*, 17 July.

Chuang, P.M. (2002c) 'S'pore wants water price pegged to NEWater's', *The Business Times*, 24 July.

Chuang, P.M. (2002d) 'KL has no right to seek water price review: S'pore', *The Business Times*, 16 October.

Chuang, P.M. (2002e) 'Onus on KL to make progress: MFA', *The Business Times*, 18 October.

Chuang, P.M. (2003f) 'M'sia wants 200-fold hike in price of raw river water', *The Business Times*, 27 January.

Chuang, P.M. (2003g) 'Singapore will strictly follow water agreements now: Jaya', *The Business Times*, 1 November.

Chuang, P.M. and Toh, E. (2003a) 'S'pore, KL sign pact to refer Pedra Blanca dispute to ICJ', *The Business Times*, 7 February.

Chuang, P.M. and Toh, E. (2003b) 'KL ad blitz a rehash of old stories: MFA', *The Business Times*, 15 July.

Chuieng, A.T. (2002) 'S'pore would agree to new water price if it's reasonable', *The Straits Times*, 16 December.

City Council of Singapore (1959) *Water Department Annual Report*, Government Printing Office, Singapore.

Clean Air Act (1971) Government Printer, Singapore.

Clean Air (Standards) Regulations (1972) S.14/1972, Government Printer, Singapore.

Cleary, G.J. (1970) *Air Pollution Control: Preliminary Assessment of Air Pollution in Singapore*, Government Printing Office, Singapore.

Code of Practice on Sewerage and Sanitary Works 2000 (2004) Singapore.

Code of Practice on Surface Water Drainage, 2000 (2006) 5th Edition, Singapore.

Crooks, Michell, Peacock and Stewart Pty Ltd. (1971) *Report. Urban Renewal and Development Project Singapore for the United Nations Development Programme (Special Fund)*. United Nations, Singapore.

Cruez, A.F. (2002) 'All issues to be tackled separately', *New Straits Times*, 17 October.

Cua, G. (1993) 'Singapore-Jakarta water pact another big boost to ties', *The Business Times*, 30 January.

Department of Statistics (2010) *Year Book of Statistics, Malaysia, 2009*, Government of Malaysia, Kuala Lumpur.

Department of Statistics, Ministry of Trade & Industry (2010) *Yearbook of Statistics*, Government of Singapore, Singapore.

Deutsche Presse-Agentur (7 July 1998) 'Malaysia will give Singapore water but with conditions'.

Deutsche Presse-Agentur (23 November 1998) 'Singapore and Malaysia hammering out water, funding details'.

Deutsche Presse-Agentur (9 June 1999) 'Singapore refutes Malaysia water profiteering claim'.

Deutsche Presse-Agentur (14 June 1999) 'Singapore looking for more sources of water'.

Dhanabalan, S. (1991) 'Foreword', in *Living the Next Lap: Towards a Tropical City of Excellence*, Urban Redevelopment Authority, Singapore, p. 3.

Dhillon, K.S. (2009) *Malaysian Foreign Policy in the Mahathir Era 1981–2003, Dilemmas of Development*, NUS Press, Singapore.

Dix, G.B. (1959) 'The Singapore Master Plan', *Earoph: News and Notes*, September, pp. 3–6. Quoted in Teo Siew Eng (1992) 'Planning principles in pre- and post-independent Singapore', *The Town Planning Review*, vol. 6, no. 2, p. 168.

Dobbs, S. (2003) *The Singapore River: A Social History 1819–2002*, Singapore University Press, Singapore.

Economic Development Board (EDB) (1991) *Singapore Briefing No. 25*, Economic Research Department, DBS Bank, Singapore.

The Economist (11 January 2003) 'Introducing Newater'.

Elias, S. (2003) 'Singapore counters ad on water deal', *The Star*, 26 July.

Employment Act (1968) Act 17 of 1968, Singapore National Printers Ltd., Singapore.

En-Lai, Y. (2000) 'Riau in Sumatra keen to fill S'pore's water needs', *The Straits Times*, 2 July.

En-Lai, Y., Hwee, L. and Ting, S.T. (2000) 'Massive water project is floated', *The Straits Times*, 2 July.

Environmental Public Health (Public Cleansing) Regulations (1970) Singapore National Printers Ltd., Singapore.

Environmental Public Health Act (EPHA) (1968) Act 32 of 1968, Government Printer, Singapore.

Environmental Public Health Act (EPHA) (1992) *Cap. 95, Amendment*, Government Printer, Singapore.

Environmental Public Health Act (EPHA) (1999) Act 22 of 1999, Government Printer, Singapore Government Gazette, Acts Supplement, 31 December 1971, Act 29 of 1971, pp. 287–300.

Esty, D.C. and Winston, A.S. (2006) *Green to Gold: How Smart Companies Use Environmental Strategy to Innovate, Create Value, and Build Competitive Advantage*, Yale University Press, New Haven.

Falle, S. (1971) *Letter to D.P. Aiers Esq, South-West Pacific Department, Foreign and Commonwealth Office, London*, FCO 24/1208, 4 March, National Archives of the United Kingdom, Richmond.

Feirn, V. (1971) *Letter to Mr. J.S. Chick, South-West Pacific Department*, FCO 24/1208, National Archives of the United Kingdom, Richmond.

Fernandez, W. (1998) 'Surprise water supply cut for 30,000 homes', *The Business Times*, 27 June.

Foo, K.B. (1993) 'Pollution control in Singapore: towards an integrated approach', in C. Briffett and L.L. Sim (eds) *Environmental Issues in Development and Conservation*, Proceedings, NUS, Singapore, pp. 29–42.

Foo, T.S. (1996) 'Urban environmental policy: the use of regulatory and economic instruments in Singapore', *Habitat International*, vol. 20, no. 1, pp. 5–22.

Fuller, T. (1998) 'Malaysia's leader fires hot words at Singapore', *International Herald Tribune*, 5 August.

George, C. (2006) 'Contentious journalism and the Internet: towards democratic discourse in Malaysia and Singapore', p. 42. Quoted in C.M. Turnbull (2009) *A History of Singapore (1819–2005)*, NUS Press, Singapore.

George, C. (2007) 'Singapore's emerging informal public sphere: a new Singapore', in T. Tan (ed.) *Singapore Perspectives 2007*, World Scientific Publishing Co., Singapore, pp. 93–104.

Ghani, A. and Hooi, A. (2005) 'MM's vision of city dam takes shape', *The Straits Times*, 23 March.

Ghesquière, H.C. (2007) *Singapore's Success: Engineering Economic Growth*, Thomson Learning, Singapore.

Goh, C.C. (1959) *Speech on 29th March, The Task Ahead: PAP's Five Year Plan (1959–1964)*, PETIR-Organ of the People's Action Party, Singapore.

Goh, C.T. (1981) *Debate on President's Address*, Parliament no. 5, Session no. 1, vol. 40, Sitting no. 5, Sitting date: 20 February, Hansard, Singapore.

Goh, C.T. (1990) 'Interview', *The International Herald Tribune*, 12 November.

Goh, K.S. (1970a) *Increase of Water Charges. Statement by the Minister of Finance*, Parliament no. 2, Session no. 1, vol. 29, Sitting no. 5, Sitting date: 27 January, Hansard, Singapore.

Goh, K.S. (1970b) *Local Government Integration (Amendment) Bill*, Parliament no. 2, Session no. 1, vol. 29, Sitting no. 7, Sitting date: 27 January, Hansard, Singapore.

Goldblum, C. (2008) 'Planning the world metropolis on an island city scale: urban innovation as a constraint and tool for global change', in T.C. Wong, B. Yuen and C. Goldblum (eds) *Spatial Planning for a Sustainable Singapore*, Springer, Dordrecht, pp. 17–29.

Gomez, L. (2009) 'Mouth-watering prospect', *New Straits Times*, 25 July.

Government Gazette (1971) Acts Supplement, 31 December, Act 29 of 1971, pp. 287–300.

Government Gazette, Acts Supplement (1975) Act 5 of 1975, pp. 21–25.

Government Gazette, Acts Supplement (1999) Act 9 of 1999, pp. 109–189.

Government Gazette, Subsidiary Legislation Supplement, 14 January 1972, S 14/72, pp. 15–19.

Government Gazette, Subsidiary Legislation Supplement, 10 March 1978, S 43/78, pp. 123–124.

Government Gazette, Subsidiary Legislation Supplement, 2 May 1980, S 127/80, p. 469.

Government of Singapore (1974) *Newspaper and Printing Presses Act*. Available at: http://statutes.agc.gov.sg/non_version/cgi-bin/cgi_retrieve.pl?actno=REVED206&doctitle=NEWSPAPER%20AND%20PRINTING%20PRESSES%20ACT%0A&date=latest&method=part (accessed on 18 September 2012).

Government of Singapore (1999) *Singapore 21: Together We Make the Difference*, Singapore 21 Committee, Government of Singapore, Singapore.

Government of Singapore (2012) *Undesirable Publications Act*. Available at: http://statutes.agc.gov.sg/aol/search/display/view.w3p;page=0;query=DocId%3A%22260443c8-a729-40e2-ac3a-a4fe673d71bb%22%20Status%3Apublished%20Depth%3A0;rec=0 (accessed on 18 September 2012).

Government of Singapore, Singapore Statistics. Available at: www.singstat.gov.sg/stats/stats.html.

Guarantee Agreement between the Government of Malaysia and the Government of the Republic of Singapore signed in Johore in 24 November 1990. Available at: http://www.mfa.gov.sg/kl/doc.html (accessed on 15 March 2011).

Guillot, X. (2008) 'Vertical living and the garden city', in T.C. Wong, B. Yuen and C. Goldblum (eds) *Spatial Planning for a Sustainable Singapore*, Springer, Dordrecht, pp. 151–167.

Gunasingham, A. (2009) 'Water-week to draw 10k delegates', *The Straits Times*, 17 April.

Habib, S. (2002) 'PM: it's fine if water pact not renewed', *The Star*, 7 August.

Habib, S. and Shari, I. (2002) 'Water price can be reviewed', *The Star*, 2 November.

Han, F.K. (2002) 'It's a watertight agreement, please', *The Straits Times*, 16 February.

Hansard (1967) *Session on Budget, Loans and General*, Parliament no. 1, Session no. 1, vol. 26, Sitting date: 20 December.

Hansard (1969a) *Increase of Water Charges*, Statement by the Minister of Finance, Parliament no. 2, Session no. 1, vol. 29, Sitting no. 1, Sitting date: 23 December, Singapore.

Hansard (1969b) Oral Answers to Questions: Floods (Preventive Measures), Parliament no. 2, Session no. 1, vol. 29, Sitting no. 6, Sitting date: 29 December, Singapore.

Hansard (1970a) Environmental Public Health (Amendment) Bill, Parliament no. 2, Session no. 1, vol. 30, Sitting no. 5, Sitting date: 2 September, Singapore.

Hansard (1970b) *Local Government Integration (Amendment) Bill*, Parliament no. 2, Session no. 1, vol. 29, Sitting no. 7, Sitting date: 27 January, Singapore.

Hansard (1975) Water Pollution Control and Drainage Bill, Parliament no. 3, Session no. 2, vol. 34, Sitting no. 15, Sitting date: 29 July, Singapore.

Hansard (1976) *Public Utilities Board. Increase in Revenue*, Parliament no. 3, Session no. 2, vol. 35, Sitting no. 5, Sitting date: 18 March, Singapore.

Hansard (1977) Debate in Budget Session, Parliament no. 4, Session no. 1, vol. 36, Sitting no. 17, Sitting date: 21 March, Singapore.

Hansard (1983) Water Pollution Control and Drainage (Amendment) Bill, Session no. 1, vol. 43, Sitting no. 3, Sitting date: 20 December, Singapore.

Hansard (1989a) Oral Answers to Questions: Fishing in Reservoirs, Parliament no. 7, Session no. 1, vol. 54, Sitting no. 4, Sitting date: 11 July, Singapore.

Hansard (1989b) Oral Answer to Questions: Water Agreement with Malaysia (Progress), Parliament no. 7, Session no. 1, vol. 54, Sitting no. 5, Sitting date: 4 August, Singapore.

Hansard (1992) *Budget, Ministry of Trade and Industry*, Parliament no. 8, Session no. 1, vol. 59, Sitting no. 10, Sitting date: 11 March, Singapore.

Hansard (1999) Bill on Environmental Pollution Control, Parliament no. 9, Session no. 1, vol. 69, Sitting no. 13, Sitting date: 11 February, Singapore.

Haron, S. (1998) 'Statements do not reflect Republic's true stance', *New Straits Times*, 4 August.

Heng, S.B. and Hamsawi, R. (2002) 'MB: Singapore's decision expected', *New Straits Times*, 8 August.

Hernandez, B.Z. and Johnston, H. (1993) 'Dirty growth', *New Internationalist*, no. 246, pp. 10–11.

Ho, K.L. (2000) 'Citizen participation and policy making in Singapore: conditions and predicaments', *Asian Survey*, vol. 40, no. 3, pp. 436–455.

Ho, R. (2000) 'End reliance on neighbours for supply', *The Straits Times*, 2 December.

Hon, J. (1990) *Tidal Fortunes A Story of Change: The Singapore River and the Kallang Basin*, Landmark Books, Singapore.

Hong, C. and Abdullah, F. (2003) 'Syed Hamid: we're unhappy with pricing but we have honoured water pacts', *New Straits Times*, 7 February.

Hoong, C.L. (2002) 'KL seeking to settle water pricing separately', *The Straits Times*, 3 July.

Housing & Development Board (HDB) (1963) *Annual Report 1960*, Government Printing Office, Housing & Development Board, Singapore.

Housing & Development Board (HDB) (1965) *50,000 Up*, HDB, Singapore. Quoted in E. Waller (2001) *Landscape Planning in Singapore*, Singapore University Press, Singapore, p. 84.

Housing & Development Board (HDB) (2001) *Annual Report*, Housing & Development Board, Singapore.

How, T.T. (2002) 'KL-S'pore talks hit snag as Malaysia changes tack; Malaysia now wants water as the issue of the Customs checkpoint to be dealt with separately, instead of as a package', *The Straits Times*, 4 September.

Hsieh, D. (2000) 'Parched Beijing looks at Singapore for water tips', *The Straits Times*, 9 December.

Huifen, C. (2005) 'Water industry thriving in Singapore', *The Business Times*, 13 September.

Huifen, C. (2006) 'New body to boost environment and water industry', *The Business Times*, 18 July.

Ibrahim, R.M. (2002) 'Price of water: Malaysia says the last word to Singapore', *New Straits Times*, 2 November.

Ibrahim, Z. (2003) 'Water row not about money; issue is Singapore's sovereignty and about honouring agreements', *The Straits Times*, 26 January.

Independent, The (6 November 1998) 'Malaysia, Singapore agree to sink differences'.

International Court of Justice (ICJ) (2008) 'Sovereignty over Pedra Branca/Pulau Batu Puteh, Middle Rocks and South Ledge' (Malaysia/Singapore), *Summary 2008/1*, ICJ, The Hague, 23 May.

Jaafar, F. (2002a) 'Discussion of water issues reaches the final say', *Utusan Malaysia*, 18 October (translated from Malay into English).

Jaafar, F. (2002b) 'Water issues: Malaysia shows more tolerance', *Utusan Malaysia*, 19 October (translated from Malay into English).

Jaafar, M.A. (2003) 'The door is closed for negotiations', *Berita Harian*, 2 August (translated from Malay into English).

Jensen, R. (1967) 'Planning, urban renewal, and housing in Singapore', *The Town Planning Review*, vol. 38, no. 2, pp. 115–131.

Johore River Water Agreement between the Johore State Government and City Council of Singapore signed in Johore in 29 September 1962. Singapore Parliamentary Debates, Official Report, vol. 75, cols. 2645–2664, Annex C, 25 January 2003.

Joshi, Y., Tortajada, C. and Biswas, A.K. (2012a) 'Cleaning of the Singapore River and Kallang Basin in Singapore: human and environmental dimensions', *Ambio* 41:777–781, DOI 10.1007/s13280-012-0279-0.

Joshi, Y., Tortajada, C. and Biswas, A.K. (2012b) 'Cleaning of the Singapore River and Kallang Basin in Singapore: economic, social and environmental dimensions', *International Journal of Water Resources Development*, iFirst Article, pp. 1–12.

Kagda, S. (2000) 'S'pore Poh Lianuying into Indon Water Project', *The Business Times*, 15 June.

Kagda, S. (2001) 'New initiatives to boost economic ties with Riau', *The Business Times*, 15 February.

Kamaruddin, M. (2002) 'Water: Singapore should be sincere with Malaysia', *Utusan Malaysia*, 5 September (translated from Malay into English).

Kandiah, J. (1998) 'Treated water may be sold to Singapore under new pact', *New Straits Times*, 12 April.

Kassim, R. (1998a) '"Convergence of views" on S'pore-M'sia cooperation', *The Business Times*, 17 November.

Kassim, Y.R. (1998b) 'KL passes up US$4b S'pore loan, ties water to other issues', *The Business Times*, 18 December.

Kassim, Y.R. (2002) 'Ending S'pore-KL "hydro-politics"', *The Business Times*, 1 August.

Kassim, Y.R. (2003a) 'Can the Tiger and Sang Kancil ever make up?', *The Business Times*, 26 February.

Kassim, Y.R. (2003b) 'From Singapore Creek to the Johore River', *The Business Times*, 7 March.

Kassim, Y.R. (2003c) 'Writing water from air: the battle hots up', *The Business Times*, 11 March.

Kau, Y.K. (1998) 'Water rates to go up to instil need to conserve: minister', *The Business Times*, 27 June

Kaur, S. (2001a) 'Soon: cheaper to desalinate sea water than to import it', *The Straits Times*, 15 March.

Kaur, S. (2001b) '30m gallons a day to drink, from the sea', *The Straits Times*, 22 March.

Kaur, S. (2001c) 'Add, multiply to meet water needs', *The Straits Times*, 26 May.

Kaur, S. (2002) 'New efforts to collect rainwater from wider area', *The Straits Times*, 6 September.

Kaur, S. (2003a) 'Singapore awards first seawater plant tender', *The Straits Times*, 20 January.

Kaur, S. (2003b) 'Overseas firms thirst for Newater's ultra-pure success', *Straits Times*, 11 March.

Kaur, S. and Hussain, Z. (2002) 'Most back move to be self-sufficient', *The Straits Times*, 27 July.

Kellner, D. (2000) 'Habermas, the public sphere and democracy: a critical intervention', in L.E. Hahn (ed.) *Perspectives on Habermas*, Chicago and La Salle, Open Court, pp. 259–538.

Kenyon, A. (2007) 'Transforming media market: the cases of Malaysia and Singapore', *Australian Journal of Emerging Technologies and Society*, vol. 5, no. 2, pp. 103–118. Available at: http://www.swin.edu.au/hosting/ijets/journal/V5N2/pdf/Article3-KENYON.pdf (accessed 23 September 2012).

Khalik, S. (2000) 'Water from the sea – for less than $1', *The Straits Times*, 19 April.

Khin, N. (2004) 'PUB launches new WaterHub facility', *The Business Times*, 11 December.

Khoo, S. (2002) 'Minister: I'm ready to talk', *The Star*, 3 November.

Khoo, S. (2003) 'Delaying tactic', *The Star*, 4 August.

Khoo, T.C. (no date) *Water: Turning Scarcity into Opportunity*, unpublished draft.

Khoo, T.C. (2008) 'Towards a global hydrohub', *Pure: Annual 2007/2008*, Public Utilities Board, Singapore, pp. 4–5.

Kim, W. (2001) 'Media and democracy in Malaysia. Media and democracy in Asia', *The Public*, vol. 8, no. 2, pp. 67–88.

Kog, Y.C. (2001) *Natural Resource Management and Environmental Security in Southeast Asia: Case Study of Clean Water Supplies in Singapore*, Institute of Defence and Strategic Studies, Singapore.

Kog, Y.C., Lim, I. and Long, S.R.J. (2002) *Beyond Vulnerability? Water in Singapore–Malaysia Relations*, edited by Kwa Chong Guan, Institute of Defence and Strategic Studies, Singapore.

Koh, J. (2003) 'Asia Environment doubles on debut', *The Business Times*, 12 December.

Koh, K.L. (1995) 'The garden city and beyond: the legal framework', in G.L. Ooi (ed.)

Environment and the City: Sharing Singapore's Experience and Future Challenge, Times Academic Press for Institute of Policy Studies, Singapore, pp. 148–170.

Koh, L.C. (2002) 'Water issue may yet leave a bad taste in the mouth, NEWater or not', *New Straits Times*, 7 August.

Koh, T. and Leong, C. (2009) 'Forget bottled water, tap water as good as it gets', *The Straits Times*, 2 July.

Korean Herald, The (9 August 2006) 'Seeing Singapore through a drop of water'.

Kuan, K.J. (1988) 'Environmental improvement in Singapore', *Ambio, East Asian Seas*, vol. 17, no. 3, pp. 233–237.

Kuar, S. (2001a) 'Soon: cheaper to desalinate seawater than import it', *The Straits Times*, 15 March.

Kuar, S. (2001b) '30m gallons a day to drink, from the sea', *The Straits Times*, 22 March.

Lam, L. (2002) 'Be sincere in resolving water price deadlock, S'pore told', *The Star*, 14 October.

Laquian, A.A. (2005) *Beyond Metropolis: The Planning and Governance of Asia's Mega-Urban Regions*, Woodrow Wilson Center Press and The Johns Hopkins University Press, Baltimore.

Latif, A. (2002) 'Water: a toast to more comfortable bilateral dealings', *The Straits Times*, 29 July.

Lau, L. (2002a) 'KL does not want to "lose out" in water deal', *The Straits Times*, 1 February.

Lau, L. (2002b) 'Water deal "unfair", but KL can't act alone', *The Straits Times*, 4 May.

Lau, L. (2003c) 'Water talks: KL to give its side of the story in a book', *The Straits Times*, 1 July.

Lau, L. (2002d) 'Mahatir talks of arbitration on water issues', *The Straits Times*, 8 September.

Lau, L. (2002e) 'Malaysia to seek legal recourse for water dispute', *The Straits Times*, 1 December.

Lawrence, C.C. and Aziz, M.A. (1995) 'Environmental protection programmes', in G.L. Ooi (ed.) *Environment and the City: Sharing Singapore's Experience and Future Challenge*, Times Academic Press for Institute of Policy Studies, Singapore, pp. 200–220.

Lee, B.H. (1978) 'Singapore—reconciling the survival ideology with the achievement concept', in L. Suryadinata (ed.) *Southeast Asian Affairs*, Institute of Southeast Asian Studies, Singapore, pp. 229–244.

Lee, E. (1991) 'Community, family, and household', in E.C.T. Chew and E. Lee (eds) *A History of Singapore*, Oxford University Press, Singapore, pp. 242–267.

Lee, H.C. (2005) 'PUB joins exclusive water research group', *The Straits Times*, 7 December.

Lee, H.H. (2003) 'Singapore proposal is not an offer, but an insult to us', *New Straits Times*, 21 July.

Lee, H.L. (1985) *Oral Answers to Questions on Water Demand (Growth Rate and Conservation)*, Parliament no. 6, Session no. 1, vol. 45, Sitting no. 17, Sitting date: 28 March, Hansard, Singapore.

Lee, H.L. (1986) *Oral Answers to Questions*, Parliament no. 6, Session no. 2, vol. 48, Sitting no. 9, Sitting date: 9 December, Hansard, Singapore.

Lee, H.L. (2007) *Prime Minister Speech*, ABC Waters Exhibition, Asian Civilisations Museum, Singapore, 8 February.

Lee, H.L. (2009) *A Lively and Liveable Singapore: Strategies for Sustainable Growth*, Ministry of the Environment and Water Resources and Ministry of National Development, Singapore.

Lee, H.L (2011) Speaking on Leadership Renewal–The Fourth Generation and Beyond, Kent Ridge Ministerial Forum, National University of Singapore (NUS), 5 April, in *Knowledge Enterprise*, NUS, Singapore, p. 5.

Lee, K.Y. (1966) *New Bearings in our Education System*, Ministry of Culture, Singapore.

Lee, K.Y. (1969) Letter to Minister, Ministry of National Development, 27 August 1969, Ref. PWD/223/53 (153 A). National Archives, Singapore.

Lee, K.Y. (1978) 'Community centres as central nervous system', in *Speeches: A Monthly Collection of Ministerial Speeches*, 1, 11, May.

Lee, K.Y. (1998) *The Singapore Story: Memoirs of Lee Kuan Yew*, Singapore Press Holdings and Marshall Cavendish Editions, Singapore.

Lee, K.Y. (2000) *From Third World to First; the Singapore Story: 1965–2000*, Singapore Press Holdings, Singapore.

Lee, K.Y. (2009) Personal interview, Singapore, 11 and 12 February.

Lee, K.Y. (2011) *Lee Kuan Yew: Hard Truths to Keep Singapore Going*, Straits Press, Singapore.

Lee, P.O. (2003) *The Water Issue between Singapore and Malaysia, No Solution in Sight?* Institute of Southeast Asian Studies, Singapore.

Lee, P.O. (2005) 'Water management issues in Singapore', paper presented at Water in Mainland Southeast Asia, 29 November–2 December, Siem Reap, Cambodia, Conference organized by the International Institute for Asian Studies (IIAS), Netherlands, and the Centre for Khmer Studies (CKS), Cambodia.

Lee, P.O. (2010) 'The four taps: water self-sufficiency in Singapore', in T. Chong (ed.) *Management of Success: Singapore Revisited*, Institute of Southeast Asian Studies, Singapore, pp. 417–439.

Lee, R. (2003a) 'KL's water ad ignores crucial facts, says Singapore; Foreign Minister says it's a rehash of an old arguments and is puzzled by the timing of the current campaign against the Republic', *The Straits Times*, 15 July.

Lee, R. (2003b) 'KL's water ad blitz ignores crucial facts, says Singapore', *The Straits Times*, 15 July.

Lee, S.H. (1999) 'Florida method may help S'pore cut water costs', *The Straits Times*, 2 June.

Lee, S.K. and Chua, S.E. (1992) *More Than a Garden City*, Parks & Recreation Department, Ministry of National Development, Singapore.

Lee, Y.H. (2008) *Waste Management and Economic Growth*, Ethos, Civil Services College, Singapore.

Leitmann, J. (2000) *Integrating the Environment in Urban Development: Singapore as a Model of Good Practice*, Urban Development Division, World Bank. Available at: http://www.ucl.ac.uk/dpu-projects/drivers_urb_change/urbenvironment/pdf_Planning/World%20Bank_Leitmann_Josef_Integrating_Environment_Singapore.pdf (accessed on 26 October 2012).

Leoi, S.L. (2003) 'Going it alone', *The Star*, 2 August.

Leong, A.P.B. (1990) *The Development of Singapore Law*, Butterworths, Singapore.

Leong, C. (2010) 'Eliminating "yuck": a simple exposition of media and social change in water reuse policies', *International Journal of Water Resources Development*, vol. 26, no. 2, pp. 111–124.

Leong, C.T. (2001) 'Hyflux links up with US firm to bid for project', *The Straits Times*, 18 September.

Leong, C.T. (2005) 'A new addition in Tuas you can drink to', *The Straits Times*, 13 September.

Leong, N. (2002) 'Water price to be backdated', *The Star*, 12 October.

Leong, N. (2003) 'PM: we'll tell our side of the story', *The Star*, 2 July.

Leong, W.K. (1999) 'Always looking out for water', *The Straits Times*, 14 June.

Lian, B. (2003) 'Singapore rapped over smear campaign', *New Straits Times*, 1 July.

Liew, H.Q. (2001) 'No land goes to waste in sewerage project', *The Business Times*, 30 January.

Lim, K. (2003a) 'KL's Eco water eyes Sesdaq listing', *The Business Times*, 30 March.

Lim, K. (2003b) 'Sinomem launches 100m share IPO to raise $41.5m', *The Business Times*, 11 June.

Lim, L. (2002a) 'Four big taps will keep water flowing', *The Straits Times*, 23 May.

Lim, L. (2002b) 'PM sees progress at KL talks but no final deal', *The Straits Times*, 17 June.

Lim, L. (2002c) 'S'pore wants water price pegged to NEWater cost', *The Straits Times*, 24 July.

Lim, L. (2002d) 'Malaysia blamed for talks breakdown', *The Straits Times*, 1 November.

Lim, L. (2002e) 'Singapore ready to go for arbitration over water issue', *The Straits Times*, 29 December.

Lim, L. (2003) 'An accord followed by a little sparring', *The Straits Times*, 7 February.

Lim, M.C. (1997) *Drainage Planning and Control in the Urban Environment. Environmental Monitoring and Assessment* 44, Kluwer Academic Publishers, the Netherlands, pp. 183–197.

Lim, R. (1998a) 'Work on desalination plant likely to begin by mid-'99', *The Business Times*, 2 March.

Lim, R. (1998b) 'PUB embarking on various projects to ensure long-term water supply', *The Business Times*, 21 August.

Ling, T.L. (1969) *Interim Report on Improvement to Rivers & Canals in Singapore*, Draft, PWD/223/53/IV/181A, SEE (D&M)/PWD, 7 November, National Archives, Singapore.

Liu, T.K. (1997) 'Towards a tropical city of excellence', in G.L. Ooi and K. Kwok (eds) *City and the State, Singapore's Built Environment Revisited*, Oxford University Press, Singapore, pp. 31–43.

Loh, D. (2003) 'NEAC rebuts Singapore water ad', *New Straits Times*, 28 July.

Loh, D. (2011a) *Singapore Water Management – Supply of New Water*, Asean Water Conference, Metropolitan Waterworks Authority of Thailand, Bangkok, 2 June.

Loh, D. (2011b) *Singapore's Experience in the Supply of NEWater through a Secondary Distribution System*, Public Utilities Board, Singapore.

Loh, H.Y. (2003) 'Asia Environment gets SGX's IPO nod', *The Business Times*, 7 November.

Lokesnikov-Jessop, S. (2006) 'Singapore taps ocean for water and income', *International Herald Tribune*, 12 September.

Long, J. (2001) 'Desecuritizing the water issue in Singapore–Malaysia relations', *Contemporary Southeast Asia*, vol. 23, no. 3, pp. 504–532.

Low, E. (2001) 'Reclaimed water to meet 20% of Singapore's needs by 2010', *The Business Times*, 26 May.

Low, E. (2002) 'KL wants to discontinue package approach on outstanding issues' *The Business Times*, 12 October.

Low, E. and Toh, E. (2002) 'KL wants to discontinue package approach on outstanding issues; Dr M wants to backdate revised price of water', *The Business Times*, 12 October.

Low, L. (1997) 'The political economy of the built environment revisited', in G.L. Ooi and K. Kwok (eds) *City and the State, Singapore's Built Environment Revisited*, Oxford University Press, Singapore, pp. 78–107.

Loy, W.S. (1986) *Phasing out of Pig and Open Range Duck Farms in Water Catchments*, COB-SEA Workshop on Cleaning-up of Urban Rivers, Ministry of the Environment Singapore and UNEP, 14–16 January, Singapore.

Luan, I.O.B. (2010) 'Singapore water management policies and practices', *International Journal of Water Resources Development*, vol. 26, no. 1, pp. 65–80.

Lye, L.H. (2008) 'A fine city in a garden: environmental law, governance and management in Singapore', *Singapore Journal of Legal Studies*, pp. 68–117.

McLoughlin, J. and Bellinger, E.G. (1993) *Environmental Pollution Control: An Introduction to Principles and Practices of Administration*, Graham & Trotman/Martinus Nijhoff, London.

Maharis, M. (2001) 'We should not depend on others to supply our basic water needs', *New Straits Times*, 17 January.

Mak, K.C. (1986) *Jurong Town Corporation: A Case Study on Resettlement of Squatters and Backyard Industries in Kampong Bugis*, COBSEA Workshop on Cleaning-up of Urban Rivers, Ministry of the Environment Singapore and UNEP, 14–16 January, Singapore.

Mattar, A. (1987) Speeches. Speech on 8 December, vol. 11, no. 6, p. 19. Ministry of Communications and Information, Singapore.

Megan, M.K. (2002) 'Yes to water price review', *New Straits Times*, 9 October.

Megan, M.K. and Chan, J. (2003) 'DPM: let's not be emotional', *New Straits Times*, 14 July.

Meng, Y.C. (2006) 'S'pore No. 1 in managing water resources: expert', *The Straits Times*, 23 August.

MFAS (Ministry of Foreign Affairs of Singapore) (2003) *Statement by Minister for Foreign Affairs, Prof. S. Jayakumar*, Singapore Government, Parliament, 25 January. Available at: http://www.mfa.gov.sg/internet/press/water/speech.html#annex (accessed on 15 March 2010).

MFAS (Ministry of Foreign Affairs of Singapore) (2005) *Joint Press Release on the Meeting between Malaysia and Singapore on the Outstanding Bilateral Issues*. Available at: http://app.mfa.gov.sg/2006/press/view_press.asp?post_id=1258 (accessed on 29 June 2010).

MICA (Ministry of Information, Communications and the Arts) (2003) *Ministerial Statement by Prof. S. Jayakumar*, Singapore Minister for Foreign Affairs in the Singapore Parliament on 25 January 2003, Ministry of Information, Communications and the Arts, Singapore, Annex A, 67–80. Available at: http://www.mfa.gov.sg/internet/press/water/speech.html#annex (accessed on 20 July 2010).

Ming, C.P (2002a) 'NEWater gets panel's clearance; 2-year study clears flow of reclaimed water into reservoirs', *The Business Times*, 12 July.

Ming, C.P. (2002b) 'NEWater purer than PUB's', *The Business Times*, 17 July.

Ming, C.P. (2002c) 'KL has no legal right to seek water price review: S'pore', *The Business Times*, 16 October.

Ministry of the Environment (ENV) (1973) *Singapore Success Story: Towards a Clean and Healthy Environment*, Ministry of the Environment, Singapore.

Ministry of the Environment (ENV) (1977) *Progress Report, Cleaning Up of Singapore River/ Kallang Basin and all Water Catchments*, MOE, PWD 223/53 Vol. VI, National Archives, Singapore.

Ministry of the Environment (ENV) (1981) Annexure of draft memo no. ENV/CF 09/77 PE1, 6 May, Singapore.

Ministry of the Environment (ENV) (1986–1993) *Annual Reports*, Ministry of the Environment, Singapore.

Ministry of the Environment (ENV) (1987) *Clean Rivers: The Cleaning up of Singapore River and Kallang Basin*, Ministry of the Environment, Singapore.

Ministry of the Environment (ENV) (1993) *The Singapore Green Plan: Action Programs*, Ministry of the Environment, Singapore.

Ministry of the Environment (ENV) (1993a) *Environmental Protection in Singapore*, Handbook, Ministry of the Environment, Singapore.

Ministry of the Environment (ENV) (1993b) *Singapore Green Plan-Action Programs*, Ministry of the Environment, Singapore.

Ministry of the Environment (ENV) (2002) *National Green Plan*, Ministry of the Environment, Singapore.

Ministry of the Environment and Water Resources (MEWR) (2006) *The Singapore Green Plan 2012*, MEWR, Singapore. Available at: http://app.mewr.gov.sg/web/Contents/Contents.aspx?ContId=1342 (accessed on 14 September 2012).

Ministry of the Environment and Water Resources (MEWR) (2012) *The Singapore Green Plan Debate on Foreign Workers*. Available at: http://www.mewr.gov.sg/sgp2012/about.htm (accessed on 4 September 2012).

Ministry of the Environment and Water Resources (MEWR) and Ministry of National Development (MND) (2009) *A Lively and Liveable Singapore: Strategies for Sustainable Growth*, Ministry of the Environment and Water Resources and Ministry of National Development, Singapore.

Ministry of the Environment and Water Resources, *Key Environment Statistics – Water Resource Management*, Singapore. Available at: http://app.mewr.gov.sg/web/Contents/contents.aspx?ContId=682 (accessed on 29 August 2012).

Ministry of Finance (1996) *Budget Speech 1996*, Singapore. Available at: http://www.mof.gov.sg/budget_1996/rebates.html (accessed on 17 September 2012).

Ministry of Finance (1997) *Budget Speech 1997*, Singapore. Available at: http://www. mof.gov.sg/budget_1997/utilitiesrates.html (accessed on 17 September 2012).

Ministry of National Development (MND) (1965) *Master Plan: First Review, Report of Survey*, Ministry of National Development, Singapore.

Ministry of National Development (MND) (1985) *Revised Master Plan: Report of Survey*, Ministry of National Development, Singapore.

Ministry of National Development (MND) (2002) *Parks and Waterbodies Plan*, Ministry of National Development, Singapore. Available at: http://www.ura.gov.sg/pwbid/ (accessed on 14 September 2012).

Ministry of National Development (MND) (2009) *An Endearing Home, A Distinctive Global City*, Ministry of National Development, Singapore.

Miyauchi, T. (2003) 'Singapore touts New Water', *The Nikkei Weekly*, 28 April.

Mohd, A., Waheed, S., Abdullah, F. and Singh, S. (2003) 'Malaysia made a scapegoat', *New Straits Times*, 31 January.

Moses, B. (2003) 'Malaysia bashing main item on Singapore menu', *New Straits Times*, 2 February.

Mulchand, A. (2008) 'Singapore's new water-policy institute has big goals', *The Straits Times*, 25 June.

Murugiah, C. (2002a) 'Singapore plans to let first of two water deals lapse in 2011', *New Straits Times*, 6 August.

Murugiah, C. (2002b) 'Malaysia to seek legal recourse if no headway made on water price issue', *New Straits Times*, 14 October.

Musa, A.G. (2002a) 'Singapore is trying to sell-out the country', *Berita Harian*, 26 October (translated from Malay into English).

Musa, A.G. (2002b) 'Recommends charging higher taxes to Singapore for water treatment plant', *Berita Harian*, 14 November (translated from Malay into English).

Mustafa, A.F. (2002) 'The price of water with Singapore should be fixed immediately', *Berita Harian*, 21 October (translated from Malay into English).

Nair, S. (2007) 'It's not simply a matter of water price', *New Straits Times*, 25 May.

Nanyang Siang Pau (6 July 2002) 'Singapore will allow Malaysians to withdraw CPF savings' (translated from Mandarin into English).

Nanyang Siang Pau (9 July 2002) 'Malaysia is waiting for Singapore to settle the water issue first' (translated from Mandarin into English).

Nanyang Siang Pau (22 July 2002) 'Malaysia has urged Singapore not to delay negotiations' (translated from Mandarin into English).

Nanyang Siang Pau (25 July 2002) 'NEWater will ease the tension of the water agreements' (translated from Mandarin into English).

Nanyang Siang Pau (2 August 2002) 'Singapore will buy water from Malaysia even though NEWater is enough' (translated from Mandarin into English).

Nanyang Siang Pau (19 October 2002a) 'Singapore is sincere in resolving the water agreement' (translated from Mandarin into English).

Nanyang Siang Pau (19 October 2002b) 'Both countries need to be patient on the water issues' (translated from Mandarin into English).

Nanyang Siang Pau (22 October 2002) 'Malaysia will not use water issues against Singapore' (translated from Mandarin into English).

Nanyang Siang Pau (24 October 2002) 'Price of water should exceed 60 cents for every 1,000 gallons' (translated from Mandarin into English).

Nanyang Siang Pau (25 October 2002a) 'Malaysia has the right to review the price of water' (translated from Mandarin into English).

Nanyang Siang Pau (25 October 2002b) 'Malaysia to pass new act to regulate water supply' (translated from Mandarin into English).

Nanyang Siang Pau (26 October 2002) 'Malaysia will study the legal aspects of the water supply to Singapore' (translated from Mandarin into English).

Nanyang Siang Pau (2 November 2002a) 'Malaysia will analyze impacts of implementing the new act' (translated from Mandarin into English).

Nanyang Siang Pau (2 November 2002b) 'Singapore is not since in solving water issues' (translated from Mandarin into English).

Nanyang Siang Pau (14 November 2002) 'Both countries should do their best to solve the water issue' (translated from Mandarin into English).

Nanyang Siang Pau (15 November 2002) 'Malaysia is waiting for the next round of meetings' (translated from Mandarin into English).

Nanyang Siang Pau (22 November 2002) 'SG is delaying discussions' (translated from Mandarin into English).

Nanyang Siang Pau (2 December 2002) 'Malaysia wants to settle the outstanding water issue through legal means' (translated from Mandarin into English).

Nanyang Siang Pau (27 December 2002) 'If the water dispute cannot be resolved between the two countries, it will be referred to a third party' (translated from Mandarin into English).

Nanyang Siang Pau (3 January 2003) 'MY will try to solve the water issues without legal means' (translated from Mandarin into English).

Nanyang Siang Pau (20 January 2003) 'Johor government should take back the water treatment plants' (translated from Mandarin into English).

Nanyang Siang Pau (31 January 2003) 'Singapore should not manipulate the issues' (translated from Mandarin into English).

Nanyang Siang Pau (3 February 2003) 'Water issues are not about pricing but sovereignty' (translated from Mandarin into English).

Nanyang Siang Pau (10 February 2003) 'By 2061 Singapore will be self-sufficient' (translated from Mandarin into English).

Nanyang Siang Pau (13 February 2003) 'Singapore has not created a deadlock' (translated from Mandarin into English).

Nanyang Siang Pau (20 February 2003) 'Malaysia will not stop supplying water to Singapore' (translated from Mandarin into English).

Nanyang Siang Pau (26 March 2003a) 'MY will refer to the ICJ' (translated from Mandarin into English).

Nanyang Siang Pau (26 March 2003b) 'Singapore's accusations have no grounds' (translated from Mandarin into English).

Nanyang Siang Pau (1 August 2003) 'Water supply system in Hong Kong and Kwangtung as an example for Malaysia and Singapore' (translated from Mandarin into English).

Nanyang Siang Pau (8 August 2003) 'How to negotiate the water supply dispute' (translated from Mandarin into English).

Nanyang Siang Pau (3 September 2003) 'No secret talks on water issues' (translated from Mandarin into English).

Nasir, A.G. (2003) 'Johor does not need clean water from Singapore', *Berita Harian*, 14 January (translated from Malay into English).

Nathan, D. (1998) 'The Sunday times', *The Straits Times*, 23 August, p. 1 PC 4.5 a, Encl. no. 4.

Nathan, D. (1999) 'New plant to test water – saving ideas', *The Straits Times*, 8 June.

Nathan, D. (2000) 'New plant sells potable water', *The Straits Times*, 3 January.

Nathan, D. (2002) 'Experts find reclaimed water safe to drink; international panel gives Singapore's NEWater thumbs up after 2-year study; nod for blending it with reservoir water', *The Straits Times*, 12 July.

National Archives of Australia (1965) *Confidential Telegram from the Secretary of State for Commonwealth Relations to the British Foreign High Commissioner*, CRS A1838/280, 3006/10/4/1 Part 1, 9 August, Canberra.

National Archives of Singapore (1953) *Government Order CSO 1163/53/21*, PWD vol. 223/53 vol. I, 9 September, National Archives, Singapore.

National Archives of Singapore (1955) *SEE (C&M)*, PWD 198/53, p. 143, 28 September, National Archives, Singapore.

National Archives of Singapore (1967) *Sim Ki Boon's* [a member of Advisory Board to the Government] *Letter to Permanent Secretary*, Ministry of National Development, PWD 223/53, vol. IV, p. 44, National Archives, Singapore.

National Archives of Singapore (1969a) *Prime Minister's Memorandum to Wong Chooi Sen*, PWD 223/53, Vol. V, p.100A, 26 March, National Archives, Singapore.

National Archives of Singapore (1969b) *Lionel de Rosario's Letter*, PWD 223/53, vol. IV, p. 167, 10 October, National Archives, Singapore.

National Archives of Singapore (1969c) Various Communications and Interim Reports, PWD 223/53, vol. IV, National Archives, Singapore.

National Archives of Singapore (1969d) *Director of Public Works' Letter*, PWD 223/53, vol. IV, p. 121, 2 June, National Archives, Singapore.

National Archives of Singapore (1969e) *Interim Report on Improvement to Rivers and Canals in Singapore*, PWD 223/53, vol. IV, p. 181A, draft submitted by Ling Teck Luke, Ag. SEE (D&M)/PWD, National Archives, Singapore.

National Archives of Singapore (1969f) *Lee Kuan Yew Letter to Minister of National Development*, PWD 223/53, p. 153A, 27 August, National Archives, Singapore.

National Archives of Singapore (1970) Various Communications and Interim Reports, PWD 223/53, Vol. V, National Archives, Singapore.

National Archives of Singapore (1971a) *Hon Sui Sen, M.P. for Havelock's letter to Director*, PWD 223/53, Vol. V, p. 117, 24 May, Singapore.

National Archives of Singapore (1971b) *Letters to Supdt. Hawker Branch*, Ministry of Health, PWD 450/6, 223/53, Vol. V, pp. 118 and 112, 25 May and 14 April, Singapore.

National Archives of Singapore (1981) *Progress Report on Cleaning Up of Singapore River/ Kallang Basin and all Water Catchments (1977–September 1981)*, Ministry of the Environment, PWD 223/53, vol. VI, p. 15A, Singapore.

National Archives of Singapore (2007) Interview with Mr. Tan Gee Paw, as part of 'The civil service – a retrospective', *Series Interview*, 6 November to 11 December, National Archives, Singapore.

National Archives of Singapore (2008) *10 Years that Shaped a Nation 1965–1975*, Singapore.

National Archives of United Kingdom (no date) *Appraisal of Singapore Sewerage Project*, Ministry of Overseas Development and Overseas Development Administration, Malaysia and Singapore Department and successors, Registered Files (MS Series), World Bank aid for Singapore for the 1967–1969 period, OD 39/105, Kew.

National Climate Change Secretariat (2012) *National Climate Change Strategy 2012*, Prime Minister's Office, Singapore.

National Environment Agency (NEA) (2008) *Annual Report, National Environment Agency, Singapore*. Available at: http://web1.env.gov.sg/cms/epd_ar2008/epd08.pdf (accessed on 7 September 2012).

National Environment Agency (NEA) (2009) *Annual Report*, National Environment Agency, Singapore.

NEAC (National Economic Action Council) (2003) *Water: The Singapore–Malaysia Dispute: The Facts*. National Economic Action Council, Kuala Lumpur. Available at: http://the-star.com.my/archives/2003/7/21/nation/waterbooklet3.pdf (accessed on 15 March 2010).

Neo, B.S. and Chen, G. (2007) *Dynamic Governance: Embedding Culture, Capabilities and Change in Singapore*, World Scientific Publishing Co. Pte. Ltd, Singapore.

New Straits Times (4 April 1998) 'KL and Singapore set to resolve water supply issues'.

New Straits Times (5 August 1998) 'Dr M: we will not cut water supply to Singapore'.

New Straits Times (8 June 1999) 'Local water needs the priority'.

New Straits Times (13 March 2000) 'Syed Hamid denies that UMNO elections holding up talks with Singapore'.

New Straits Times (10 January 2001) 'Present water treaty not in favour of Malaysia'.

New Straits Times (23 January 2002) 'Settle water issue quickly'.

New Straits Times (26 January 2002) 'DPM: priority for resolving water issue'.

New Straits Times (31 January 2002) 'Johor to stop buying water from Singapore'.

New Straits Times (7 April 2002) 'Realism in diplomacy'.

New Straits Times (22 June 2002) 'Dr. M: we are prepared to take over'.

New Straits Times (3 July 2002) 'A cheap shot at a cheap price'.

New Straits Times (6 August 2002) 'Singapore plans to let first of two water deals lapse in 2011'.

New Straits Times (8 August 2002) 'End it now, being anew'.

New Straits Times (14 October 2002) 'We want to resolve matter, but not at all cost'.

New Straits Times (20 October 2002) 'Ghani: take water issue to court if talks end in deadlock'.

New Straits Times (21 October 2002l) 'Singapore must decide, says Ghani'.

New Straits Times (24 October 2002) 'Singapore reluctant to solve water issue'.

New Straits Times (1 November 2002) 'Singapore to contest water price review'.

New Straits Times (4 November 2002) 'Syed Hamid on water issue queries'.

New Straits Times (11 November 2002) 'Concerned citizen urges Singapore to renegotiate water price with Malaysia'.

New Straits Times (14 November 2002) 'Singapore. Water talks back to square one'.

New Straits Times (18 November 2002) 'Malaysia sees good signs in Singapore's willingness to talk'.

New Straits Times (20 November 2002) 'Singapore should be thankful to Johor'.

New Straits Times (22 November 2002) 'Singapore's waver a trick to delay water price negotiations, says MB'.

New Straits Times (30 December 2002) 'No reason to take water issue to arbitration'.

New Straits Times (2 January 2003) 'We need to get though with Singapore'.

New Straits Times (9 January 2003) 'Plan to stop buying treated water due to plant completion', *New Straits Times*.

New Straits Times (30 January 2003) 'Ghani: water supply to Singapore to stop after 2061'.

New Straits Times (1 February 2003) 'Singapore: water dispute is not about money'.

New Straits Times (2 February 2003) 'Don't blow your top'.

New Straits Times (10 February 2003) 'Water: Singapore to be self-sufficient by 2061'.

New Straits Times (25 February 2003) 'Singapore's overdoing it with personal attacks'.

New Straits Times (1 July 2003) 'Assemblyman criticises hike in water tariffs'.

New Straits Times (18 July 2003a) 'Singapore implies that Malaysia is mean and miserly in water deal'.

New Straits Times (18 July 2003b) 'Water advertisements: facts are new to many people, says PM'.

New Straits Times (19 July 2003a) 'NEAC: asking for fair price is not bullying'.

New Straits Times (19 July 2003b) 'Scrounging on Malaysian goodwill'.

New Straits Times (21 July 2003) 'Is fair price too much to ask?'.

New Straits Times (22 July 2003) 'Booklet on water dispute selling like hot cakes'.

New Straits Times (26 July 2003a) 'Niz Aziz: government did the right thing in informing people'.

New Straits Times (26 July 2003b) 'Malaysia to counter again Singapore's ad in AWSJ'.

New Straits Times (5 August 2003) 'Onward to arbitration'.

New Straits Times (13 September 2003) 'PM: Singapore testing our patience with baseless accusations'.

New Straits Times (14 April 2006) 'Plan wouldn´t have worked anyway'.

New Straits Times (16 July 2006) 'What they wrote in the four letters'.

Newsweek (16 July 2001) 'Not any drop to drink'.

Ng, I. (2001a) 'Tough talks, then progress on KL pact', *The Straits Times*, 5 September.

Ng, I. (2001b) 'S'pore-KL pact's success hinges on details', *The Straits Times*, 15 September.

Ng, I. and Pereira, B. (2001) 'Thorny issues that go back many years', *The Straits Times*, 5 September.

Ng, K.H. (1998) 'Overview of water conservation in Singapore', in Economic & Social

Commission for Asia & the Pacific, *Towards Efficient Water Use in Urban Areas in Asia and the Pacific*, United Nations, New York, pp. 32–37.

Ng, L. (2008) *A City in a Garden*, Ethos, World Cities Summit Issue, Civil Services College, Singapore, pp. 62–68.

Ng, L. and Oon, Y. (1998) 'Disputes talking toll on ties between Singapore, Malaysia', *The Nikkei Weekly*, 20 July.

Ng, P. (2006) 'KL declassifies confidential documents on S'pore issues', *The Business Times*, 15 July.

Nikkei Weekly, The (22 May 2006) '"Water queen" crowned for treatment business'.

Noma, K. (2008) 'Singapore immersing itself in advanced water technologies', *The Nikkei Weekly*, 8 September.

OECD (Organization for Economic Co-operation and Development) (2008) *Cost of Inaction on Key Environmental Challenges*, Organization for Economic Co-operation and Development, Paris.

Omar, A. (2002) 'Of water supply and Singapore smart talk', *New Straits Times*, 8 August.

Ong, C. (1998) 'S'pore, KL have not reached accord on water', *The Business Times*, 30 June.

Ong, E.G. (1959a) *Parliamentary Report on City Council (Suspension and Transfer of Functions) Bill*, Parliament no. 1, Session no. 1, vol. 11, Sitting no. 3, Sitting date: 16 July, Hansard, Singapore.

Ong, E.G. (1959b) *Speech on 21 February, The Task Ahead: PAP's Five Year Plan (1959–1964)*, PETIR-Organ of the People's Action Party, Singapore, pp. 29–31.

Ooi, G.L. (no date) As You Drink, Remember the Source, Singapore (unpublished work).

Ooi, G.L. (2002) 'The role of the state in nature conservation in Singapore', *Society & Natural Resources*, vol. 15, no. 5, pp. 455–460.

Ooi, G.L (2005) *Sustainability and Cities: Concept and Assessment*, World Scientific, Singapore.

Oon, Y. (2001) 'Summit yields significant progress', *The Nikkei Weekly*, 17 September.

Oorjitham, S. (2002) 'Reclamation affects Johor ports', *New Straits Times*, 18 March.

Oriental Daily News (16 July 2003) 'Singapore should be responsible for the media war'.

Osman, M. (2002a) 'Malaysia still willing to pump millions of gallons of scarce water to Singapore', *New Straits Times*, 2 September.

Osman, M. (2002b) 'Tough first day at Singapore water talks', *New Straits Times*, 3 September.

Osman, M. (2002c) 'Singapore agrees to discuss price', *New Straits Times*, 4 September.

Osman, A. (2004) 'Abdullah opens water treatment plant in Johor', *The Straits Times*, 13 February.

Parliamentary Debates (1971) *Addendum to Presidential Address*, Official Reports, vol. 31, 21 July, National Printers, Singapore.

Parliamentary Debates (1999) Debate on Environment Pollution Control Bill, Parliament no. 9, Session 1, vol. 69, Sitting 13, Sitting date: 11 February, Hansard, Singapore.

Peh, S.H. (2007) 'Fourth Newater plant to open next week', *The Straits Times*, 7 March.

People's Action Party (PAP) (1959) *The Task Ahead: PAP's Five Year Plan (1959–1964)*, PETIR-Organ of the PAP, Singapore.

Pereira, B. (2002a) 'Mahatir ready to end water pact', *The Straits Times*, 7 August.

Pereira, B. (2002b) 'Water issue: KL believes S'pore is softening stand', *The Straits Times*, 19 November.

Pereira, B. (2002c) 'KL insists it will discuss only water-price review', *The Straits Times*, 20 November.

Pereira, B. (2003a) 'Water price not befitting S'pore, says Johor chief', *The Straits Times*, 20 January.

Pereira, B. (2003b) 'Singapore stance on water gets scant coverage', *The Straits Times*, 27 January.

Pereira, B. (2003c) 'Mahathir dismisses sovereignty issue and all talk of war', *The Straits Times*, 31 January.

Pereira, D. (2000) 'S'pore wants only profits, says Gus Dur', *The Straits Times*, 27 November.

Pereira, D. (2001) 'Riau-S'pore ties to go beyond water, greens', *The Straits Times*, 15 February.

Pereira, B. and Lim, L. (2002) 'KL no longer wants to settle issues as package; Mahathir states this in a letter to PM Goh; he also wants to backdate any price hike of water to 1986 and 1987', *The Straits Times*, 12 October.

Perry, M., Kong, L. and Yeoh, B.S.A. (1997) *Singapore: A Developmental City-State*, John Wiley, Chichester.

Phan, M. (2005) 'UEL forms unit to seek regional water-treatment projects', *The Business Times*, 28 December.

Phan, M. (2006) 'S'pore resource mgt can be a model for others: panel', *The Business Times*, 17 November.

Pilihan, S. (2003) 'Singapore finds ways to hurt its neighbours', *Berita Harian*, 1 July (translated from Malay into English).

Pollution Control Department (PCD) (1993) *1992 Pollution Control Report*, Pollution Control Department, Ministry of the Environment, Singapore.

Pollution Control Department (PCD) (1994) *1993 Pollution Control Report*, Pollution Control Department, Ministry of the Environment, Singapore.

Pollution Control Department (PCD) (1997) *Code of Practice on Pollution Control*, Ministry of the Environment, Singapore.

Pollution Control Department (PCD) (2007) *Annual Report*, Ministry of the Environment, Singapore.

Pollution Control Department (PCD) (2008) *Annual Report*, Ministry of the Environment, Singapore.

Pollution Control Department (PCD) (2009) *Annual Report*, Ministry of the Environment, Singapore.

Poon, I.H. (1986) *PSA's Role in the Cleaning Up Programme*, COBSEA Workshop on Cleaning-up of Urban Rivers, Ministry of the Environment Singapore and UNEP, 14–16 January, Singapore.

Prohibition on Smoking in Certain Places Act (1970) Act 26 of 1970, Lim Bian Han Government Printer, Singapore.

Puah, A.K. (2011) *Smart Water – Singapore Case Study*, Smart Water Cluster Workshop, IWA-ASPIRE Conference, Tokyo, 2 October.

Public Utilities Act (2001) Act 8 of 2001, Government Printer, Singapore.

Public Utilities Board (PUB) (1963) *Annual Report*, PUB, Singapore.

Public Utilities Board (PUB) (1964) *Annual Report*, PUB, Singapore.

Public Utilities Board (PUB) (1965) *Annual Report*, PUB, Singapore.

Public Utilities Board (PUB) (1966) *Annual Report*, PUB, Singapore.

Public Utilities Board (PUB) (1967) *Annual Report*, PUB, Singapore.

Public Utilities Board (PUB) (1968) *Annual Report*, PUB, Singapore.

Public Utilities Board (PUB) (1969) *Annual Report*, PUB, Singapore.
Public Utilities Board (PUB) (1970) *Annual Report*, PUB, Singapore.
Public Utilities Board (PUB) (1971) *Annual Report*, PUB, Singapore.
Public Utilities Board (PUB) (1972) *Annual Report*, PUB, Singapore.
Public Utilities Board (PUB) (1973) *Annual Report*, PUB, Singapore.
Public Utilities Board (PUB) (1975) *Annual Report*, PUB, Singapore.
Public Utilities Board (PUB) (1976) *Annual Report*, PUB, Singapore.
Public Utilities Board (PUB) (1977) *Annual Report*, PUB, Singapore.
Public Utilities Board (PUB) (1978) *Annual Report*, PUB, Singapore.
Public Utilities Board (PUB) (1979) *Annual Report*, PUB, Singapore.
Public Utilities Board (PUB) (1980) *Annual Report*, PUB, Singapore.
Public Utilities Board (PUB) (1981) *Annual Report*, PUB, Singapore.
Public Utilities Board (PUB) (1983) *Annual Report*, PUB, Singapore.
Public Utilities Board (PUB) (1985a) *Annual Report*, PUB, Singapore.
Public Utilities Board (PUB) (1985b) *Yesterday & Today: The Story of Public Electricity, Water and Gas Supplies in Singapore*, PUB, Singapore.
Public Utilities Board (PUB) (1986) *Annual Report*, PUB, Singapore.
Public Utilities Board (PUB) (1987) *Annual Report*, PUB, Singapore.
Public Utilities Board (PUB) (1988) *Annual Report*, PUB, Singapore.
Public Utilities Board (PUB) (1989) *Annual Report*, PUB, Singapore.
Public Utilities Board (PUB) (1990) *Annual Report*, PUB, Singapore.
Public Utilities Board (PUB) (1992) *Annual Report*, PUB, Singapore.
Public Utilities Board (PUB) (1993) *Annual Report*, PUB, Singapore.
Public Utilities Board (PUB) (1994) *Annual Report*, PUB, Singapore.
Public Utilities Board (PUB) (1995) *Annual Report*, PUB, Singapore.
Public Utilities Board (PUB) (1996) *Annual Report*, PUB, Singapore.
Public Utilities Board (PUB) (1997a) *Annual Report*, PUB, Singapore.
Public Utilities Board (PUB) (1997b) *Singapore Government Press Statement*, PUB, Singapore.
Public Utilities Board (PUB) (1998a) *Circular on the Mandatory Water Efficiency Labelling Scheme (MWELS) and Mandatory Installation of Dual Flush Low Capacity Flushing Cisterns*, PUB, Singapore, 23 October. Available at: http://www.ies.org.sg/e-newsletter/PUBMWELS.pdf (accessed on 14 September 2012).
Public Utilities Board (PUB) (1998b) *Annual Report*, PUB, Singapore.
Public Utilities Board (PUB) (1999) *Annual Report*, PUB, Singapore.
Public Utilities Board (PUB) (2001) *Annual Report*, PUB, Singapore.
Public Utilities Board (PUB) (2002a) *Annual Report*, PUB, Singapore.
Public Utilities Board (PUB) (2002b) *Water: Precious Resource for Singapore*, PUB, Singapore.
Public Utilities Board (PUB) (2003) *Annual Report*, PUB, Singapore.
Public Utilities Board (PUB) (2004) *Water for All: Annual Report*, PUB, Singapore.
Public Utilities Board (PUB) (2005a) *Towards Environmental Sustainability: State of the Environment 2005 Report*, PUB, Singapore.
Public Utilities Board (PUB) (2005b) *Water for All: Annual Report*, PUB, Singapore.
Public Utilities Board (PUB) (2006) *Annual Report*, PUB, Singapore.
Public Utilities Board (PUB) (2006–2007) *Water for All: Annual Report*, PUB, Singapore.
Public Utilities Board (PUB) (2008a) *Annual Report*, PUB, Singapore.
Public Utilities Board (PUB) (2008b) Correspondence to the President, Institution of Engineers, Singapore, on the Mandatory Water Efficiency Labelling Scheme (MWELS)

and the Mandatory installation of Dual Flush Low Capacity Flushing Cisterns, PUB, Singapore, 23 October.

Public Utilities Board (PUB) (2010) *Water for All: Conserve, Value, Enjoy*, PUB, Singapore.

Public Utilities Board (PUB) (2011a) 3P Approach Analysis, Public Utilities Board, Singapore.

Public Utilities Board (PUB) (2011b) *Water for All: Meeting Our Water Needs for the Next 50 Years*, PUB, Singapore.

Public Utilities Board (PUB) (2012) *Innovation in Water Singapore*, PUB, Singapore.

Public Utilities Ordinance (1963) Act 1 of 1963, Lim Bian Han Acting Government Printer, Singapore.

Quah, J.S.T. (1991) 'The 1980s: A Review of Significant Political Developments', in E.C.T. Chew and E. Lee (eds) *A History of Singapore*, Oxford University Press, Singapore, pp. 385–400.

Quah, J.S.T. (2010) *Public Administration Singapore Style*, Emerald Group, Bingley.

Quah, S.R. (1983) 'Social discipline in Singapore: an alternative for the resolution of social problems', *Journal of Southeast Asian Studies*, vol. 14, no. 2, pp. 266–289.

Raaff, A.R. (2003) 'Johor should treat its water', *Berita Harian*, 10 January (translated from Malay into English).

Ramachandran, S. (2007) 'Learn how to manage water resources', *New Straits Times*, 18 November.

Rock, M.T. (2002) *Pollution Control in East Asia: Lessons from the Newly Industrializing Economies*, Resources for the Future, Washington, DC.

Said, R. (2002) 'Discussion with Singapore stalled due to water issue', *New Straits Times*, 22 January.

Said, R. (2003a) 'Singapore action criticized (HL)', *New Straits Times*, 28 January.

Said, R. (2003b) 'Singapore RM 662m profit', *New Straits Times*, 13 July.

Said, R. (2003c) 'Only a few cents for water', *New Straits Times*, 15 July.

Said, R. (2003d) 'No agreement on water price', *New Straits Times*, 17 July.

Said, R. (2003e) 'Dr M not confident of water pact', *New Straits Times*, 2 July.

Said, R. and Cruez, F. (2003) 'Malaysia has right to review', *New Straits Times*, 16 July.

Said, R. and Darshni, S. (2002) 'PM: No more cheap water', *New Straits Times*, 12 October.

Said, R. and Loh, D. (2002) 'Set time frame for water dispute', *New Straits Times*, 7 September.

Said, S. (2002) 'S'pore will not renew 1961 water agreement, says Goh', *Bernama*, 5 August.

Sale of Food Act (Amendment) (2002) Act 7 of 2002, Government Printer, Singapore.

Sanitary Appliances and Water Charges Regulations (1975) SLS 43/75, Singapore National Printers, Singapore.

Savage, V.R. (1991) 'Singapore garden city: reality, symbol, ideal', *Solidarity*, pp. 131–132.

Saw, S.H. (1991) 'Population growth and control', in E.C.T. Chew and E. Lee (eds) *A History of Singapore*, Oxford University Press, Singapore, pp. 219–241.

Saw, S.H. and Kesavapany, K. (2006) *Singapore–Malaysia Relations under Abdullah Badawi*, Institute of Southeast Asian Studies, Singapore.

Sayuthi, S. (2002) 'Singapore was ready to go to war', *New Straits Times*, 8 April.

Science Council of Singapore (1980) *Environmental Protection in Singapore*, Handbook, Singapore.

Seetharam, K. and Araral, E. (2008) 'Bursting 6 myths about water governance', *The Straits Times*, 10 July.

Seetoh, K.C. and Ong, A.H.F. (2008) 'Achieving sustainable industrial development through a system of strategic planning and implementation: the Singapore model', in T.C. Wong, B. Yuen and C. Goldblum (eds) *Spatial Planning for a Sustainable Singapore*, Springer, Dordrecht, pp. 113–133.

Seong, C.C. (2002) 'MB: Singapore's stand on water issue a stumbling block to ties', *New Straits Times*, 8 September.

Sewage Treatment Plants (Amendment) Regulations (1978) SLS 26/78, Singapore National Printers, Singapore.

Sewerage and Drainage Act (Chapter 294), SNP Corporation, Singapore.

Sewerage and Drainage Act (SDA) (1999) Act 10 of 1999, Government Printer, Singapore.

Shiraishi, T. (2009) *Across the Causeway: A Multidimensional Study of Malaysia-Singapore Relations*, Institute of Southeast Asian Studies, Singapore.

Shoichiro, H. (1991) *Water as Environmental Art: Creating Amenity Space*, Kashiwashobo Publishing Co. Ltd., Tokyo.

Shriver, R. (2003) *Malaysian Media: Ownership Control and Political Content*. Available at: http://www.rickshriver.net/Documents/Malaysian%20Media%20Paper%20-%20CAR-FAX2.pdf (accessed on 3 April 2010).

Sidhu, J.S. (2006) 'Malaysia–Singapore relations since 1998: a troubled past—whither a brighter future?', in R. Harun (ed.) *Malaysia's Foreign Relations: Issues and Challenges*, University Malaya Press, Kuala Lumpur, pp. 75–92.

Sim, S. (1999) 'Water deal with Jakarta is possible, says Philip Yeo', *The Straits Times*, 16 January.

Sim, S. (2000) 'The outburst from Gus Dur', *The Straits Times*, 30 November.

Sin Chew Daily (4 July 2002) 'Malaysia is willing to supply water to Singapore for another 100 years, but the price will be revised' (translated into English from Mandarin).

Sin Chew Daily (2 August 2002) 'Water supply must be based on mutual agreement' (translated into English from Mandarin).

Sin Chew Daily (2 September 2002) 'SG Deputy Prime Minister hopes that the second round of hopes will see some progress' (translated into English from Mandarin).

Sin Chew Daily (19 October 2002) 'Malaysia may not have been sincere on resolving water issues' (translated into English from Mandarin).

Sin Chew Daily (21 October 2002) 'Singapore must take a decision' (translated into English from Mandarin).

Sin Chew Daily (22 October 2002) 'Malaysia will not use water against Singapore' (translated into English from Mandarin).

Sin Chew Daily (25 October 2002a) 'Malaysia has the right to restrict the export of water to Singapore' (translated into English from Mandarin).

Sin Chew Daily (25 October 2002b) 'Malaysia has its right to review water price' (translated into English from Mandarin).

Sin Chew Daily (26 October 2002) 'Malaysia should study the legal aspects of the water agreements' (translated into English from Mandarin).

Sin Chew Daily (2 November 2002a) 'MY press has distorted the message of PM Goh' (translated into English from Mandarin).

Sin Chew Daily (2 November 2002) 'Singapore not since in solving water issues' (translated into English from Mandarin).

Sin Chew Daily (3 November 2002) 'Solving the water issue could be more complicated if taken to court' (translated into English from Mandarin).

Sin Chew Daily (14 November 2002) 'State should be firm in handling water issues' (translated into English from Mandarin).

Sin Chew Daily (15 November 2002) 'Malaysia waiting for Singapore to fix the next meeting' (translated into English from Mandarin).

Sin Chew Daily (18 November 2002) 'Malaysia wants to negotiate the price of water' (translated into English from Mandarin).

Sin Chew Daily (21 November 2002) 'Malaysia will only discuss water supply issues until 2061' (translated into English from Mandarin).

Sin Chew Daily (2 December 2002a) 'Malaysia wants to settle the water issue legally' (translated into English from Mandarin).

Sin Chew Daily (2 December 2002b) 'Singapore has not received any formal notice for the next round of meetings' (translated into English from Mandarin).

Sin Chew Daily (27 December 2002) 'Water supply to Singapore will not be cut-off' (translated into English from Mandarin).

Sin Chew Daily (8 January 2003) 'Treated water: do not misinterpret the intention of Johor' (translated into English from Mandarin).

Sin Chew Daily (20 January 2003a) 'The neighbour countries will resolve their differences' (translated into English from Mandarin).

Sin Chew Daily (20 January 2003b) 'Topics between Singapore and Malaysia awaiting governments to be resolved'.

Sin Chew Daily (21 January 2003) 'Johor benefits more from the sale of water to Singapore' (translated into English from Mandarin).

Sin Chew Daily (29 January 2003a) 'Singapore statements do not represent the reality' (translated into English from Mandarin).

Sin Chew Daily (29 January 2003b) 'The water negotiations demands are misleading: not reflecting true stories' (translated into English from Mandarin).

Sin Chew Daily (3 February 2003) 'The issue is not sovereignty but pricing: Johor's MB' (translated into English from Mandarin).

Sin Chew Daily (20 February 2003) 'Malaysia will always supply water to Singapore' (translated into English from Mandarin).

Sin Chew Daily (1 March 2003) 'If there is no agreement, Malaysia will settle the case in the world court' (translated into English from Mandarin).

Sin Chew Daily (26 March 2003) 'Negotiations have reached a deadlock' (translated into English from Mandarin).

Sin Chew Daily (24 July 2003) 'Singapore should pay a reasonable price for water' (translated into English from Mandarin).

Sin Chew Daily (26 July 2003a) 'The advertisement by SG is misleading' (translated into English from Mandarin).

Sin Chew Daily (26 July 2003b) 'SG Foreign Minister does not understand comments' (translated into English from Mandarin).

Sin Chew Daily (3 August 2003) 'The ICJ will take the final decision' (translated into English from Mandarin).

Sin Chew Daily (4 August 2003) 'PM says that Singapore will delay the water negotiations' (translated into English from Mandarin).

Singapore Department of Statistics (2001) *Key Annual Indicators.* Available at: http://www.singstat.gov.sg/stats/keyind.html#keyind (accessed on 24 May 2012).

Singapore Department of Statistics (2008) *Report on Household Expenditure Survey, List of Statistical Tables*, Table 16A: Average Monthly Household Expenditure by Type of Goods and Services (Detailed) and Income Quintile, Singapore.

Singapore Department of Statistics (2010) *Census of Population 2010: Advance Census Release.* Available at: http://www.singstat.gov.sg/pubn/popn/c2010acr.pdf (accessed on 24 May 2012).

Singapore Press Holdings (SPH) (2010) *Singapore Press Holdings Annual Report 2009: Growing with the Times.* Available at: http://www.sph.com.sg/annual_report.shtml (accessed on 15 January 2011).

Singapore Tourism Board (STB) (1996) *Tourism 21: Vision of a Tourism Capital*, Singapore Tourism Board, Singapore.

Singh, P. and Sennyah, P (1999) 'PM: Singapore will get only treated water after 2061', *New Straits Times*, 8 June.

Soh, T.K. (1995) 'S'pore may tap reserves to build desalination plants: Hng Kiang', *Business Times*, 17 April.

Sooi, C.C. (2002a) 'Malaysia just as eager to solve bilateral issues', *New Straits Times*, 29 January.

Sooi, C.C. (2002b) 'Singapore: send protest note if reclamation work affects deep water line', *New Straits Times*, 12 March.

Sreenivasan, V. (1993) 'PUB to spend $1.4b next year on development projects', *The Business Times*, 25 December.

Srinivasan, V. (1997) 'Self-sufficiency in water possible, but costly: BG Lee', *The Business Times*, 11 June.

Star, The (7 August 2002) 'PM: it's fine if water pact not renewed'.

Star, The (8 August 2002a) 'Johor to stop buying treated water from S'pore'.

Star, The (8 August 2002a) 'Flood of jokes over NEWater'.

Star, The (18 August 2002) 'Minister: NEWater won't affect our stand'.

Star, The (9 October 2002) 'Singapore wants to know basis for new price of water'.

Star, The (9 November 2002a) 'House rejects motion on water issue'.

Star, The (9 November 2002b) 'Review ties with Singapore'. Letters/Opinion.

Star, The (11 November 2002) 'Resolve water issue peacefully'.

Star, The (18 November 2002) 'Water talks set to resume early next .

Star, The (22 November 2002) 'Singapore dilly-dallying on water issue, says MB'.

Star, The (1 February 2003) 'It's not about money, reiterates Singapore'.

Star, The (1 July 2003) 'Book on water negotiations'.

Star, The (17 July 2003c) 'Mustapa: water ads achieved their objective'.

Statutory Corporations (Capital Contribution) (2002) Act 5 of 2002, Government Printer, Singapore.

Straits Times, The (18 August 1966) 'Two currencies show'.

Straits Times, The (22 October 1966) 'Water and power rates are up next month'.

Straits Times, The (11 December 1966) 'Two dollars day'.

Straits Times, The (9 May 1971) 'All is set for water cuts'.

Straits Times, The (10 May 1971) 'Water cuts plan shocks S'pore'.

Straits Times, The (12 May 1971) 'Cutting down on water consumption by compulsion: it's the only way to stop wastage'.

Straits Times, The (16 May 1971) 'Setback in efforts to save water'.

Straits Times, The (18 May 1971) 'Drop the three million gallons on Sunday'.

Straits Times, The (25 May 1971) 'Consumption below the 100 million mark'.

Straits Times, The (28 May 1971) 'PUB warns of action against water wastage'.

Straits Times, The (23 August 1971) 'Clean and green drive can help solve water problem'.

Straits Times, The (18 February 1976) '"Don't waste water" letters to users'.

Straits Times, The (4 April 1980) 'Top civil servants to attend RC meetings'.

Straits Times, The (9 August 1980) 'Lighters must ship out for new city'.

Straits Times, The (25 July 1981) 'Revision of water rates to reduce wastage'.

Straits Times, The (30 July 1981) 'Carrot and stick way to save water'.

Straits Times, The (22 March 1990) 'The long queues in the sixties'.

Straits Times, The (7 April 1990) 'Curbs on water usage soon if call is unheeded'.

Straits Times, The (9 April 1990) 'View water as a vital commodity, BG Yeo urges'.

Straits Times, The (21 January 1999) 'Cheaper to buy water from Johor, MB tells Singapore'.

Straits Times, The (3 March 1999) 'Malaysia can choose not to buy treated water'.

Straits Times, The (8 June 1999) 'S'pore: KL report gives a distorted picture'.

Straits Times, The (1 October 1999) 'Pahang wants to sell water to boost coffers'.

Straits Times, The (2 July 2000) 'Water: Indonesian sources'.

Straits Times, The (9 July 2000) 'New system, better water quality'.

Straits Times, The (18 September 2000) 'Politics drives need for new Johor water plant'.

Straits Times, The (2 December 2000) 'End reliance on neighbors for supply'.

Straits Times, The (19 April 2001) 'New Singapore consulate in Riau'.

Straits Times, The (7 September 2001) 'WP welcomes new water deal'.

Straits Times, The (9 October 2001) 'Use water as a weapon in S'pore ties: KL article'.

Straits Times, The (26 January 2002) 'Malaysia to name water price before talks start'.

Straits Times, The (27 January 2002) 'Singapore subsidises treated water sold to Johor'.

Straits Times, The (5 April 2002) 'Just the fact, please'.

Straits Times, The (6 April 2002) 'High time for a new approach to water'.

Straits Times, The (30 April 2002) 'Pahang wants KL to handle its water pact with S'pore'.

Straits Times, The (17 May 2002) 'Several MPs raise concerns on bilateral issues'.

Straits Times, The (24 July 2002) 'Bilateral issues resolved only as a package'.

Straits Times, The (11 August 2002) 'Johor leads pokes fun at NEWater'.

Straits Times, The (3 October 2002) 'Sell sewage to S'pore instead, says MP'.

Straits Times, The (11 October 2002) 'KL presses Singapore to resolve water issue first'.

Straits Times, The (16 October 2002) 'S'pore concessions for a water deal are off'.

Straits Times, The (23 October 2002) 'S'pore shows no desire to reach deal on water'.

Straits Times, The (26 October 2002) 'S'pore to KL: remember your promises'.

Straits Times, The (8 November 2002) 'Malaysians urge KL to let water pacts lapse'.

Straits Times, The (16 November 2002) 'Talks on price of water will continue, says KL'.

Straits Times, The (21 November 2002) 'Singapore restates stand on water talks; both the current price and future water supply should be discussed, Republic says in response to KL's latest remarks'.

Straits Times, The (1 December 2002) 'S'pore "waiting for KL's clarification on water talks"'.

Straits Times, The (11 December 2002) 'S'pore must accept review right'.

Straits Times, The (28 December 2002) 'Malaysia "serious" about arbitration over water issue'.

Straits Times, The (20 January 2003) 'What the paper didn't say'.

Straits Times, The (26 January 2003a) 'Dear Kuan Yew'.

Straits Times, The (26 January 2003b) 'Letters tell the true story'.

Straits Times, The (26 January 2003c) 'What is at stake: our very existence as a nation'.

Straits Times, The (28 January 2003) 'Dear Mahathir'.

Straits Times, The (1 February 2003) 'The real issue is not price of water'.

Straits Times, The (3 February 2003) 'Price not fixed at random'.

Straits Times, The (1 March 2003) '100% safe drinking water'.

Straits Times, The (26 March 2003) 'KL ready for talks on treated water'.

Straits Times, The (8 July 2003) 'Malaysia still keen on water talks'.

Straits Times, The (20 July 2003) 'KL has not closed door on talks'.

Straits Times, The (23 July 2003) 'Media blitz waste of money'.

Straits Times, The (26 July 2003) 'S'pore ad sets out facts on water row'.

Straits Times, The (2 August 2003) 'Water supply deal will remain: Mahathir; Federal control over resources will not affect S'pore supplies, he says, but adds that time for talks is over'.

Straits Times, The (5 August 2003) 'PAS: water dispute with Singapore a diversion'.

Straits Times, The (16 September 2005) 'PUB to expand 3 Newater facilities'.

Straits Times, The (15 November 2008) 'My point'.

Sung, T.T. (1972) 'Water resources planning and development Singapore', paper no. I/2, pp. 17–25, in *Technical Papers Accepted for the Regional Workshop on Water Resources, Environment and National Development*, Science Council of Singapore, Singapore, 13–17 March.

Tan, A. (2002a) 'HK's high raw water price includes infrastructure costs; Johor has not borne any such costs for S'pore supply: MFA', *The Business Times*, 4 February.

Tan, A. (2002b) 'MFA: S'pore didn't cut water price to 12 sen', *The Business Times*, 24 October.

Tan, A. (2003) 'Reservoirs getting NEWater from yesterday', *The Business Times*, 22 February.

Tan, C. (1990) 'Rationing common in the 60s', *The Straits Times*, 6 April.

Tan, C.K. (2001) 'Singapore should consider other sources of water supply', *The Straits Times*, 11 September.

Tan, G.P. (2009) 'The face behind Singapore's master plan', *Pure: Annual Report 2008/2009*, Public Utilities Board, Singapore.

Tan, K.L.T. and Wong, T.C. (2008) 'Public housing in Singapore: a sustainable housing form and development', in T.C. Wong, B. Yuen and C. Goldblum (eds) *Spatial Planning for a Sustainable Singapore*, Springer, Dordrecht, pp. 135–150.

Tan, S. (1999) 'Johor also benefits from water agreement', *New Straits Times*, 17 June.

Tan, S.A. (1972) 'Urban renewal in Singapore and its associated problems', paper no. III/3, pp. 334–335, in *Technical Papers Accepted for the Regional Workshop on Water Resources, Environment and National Development*, Science Council of Singapore, Singapore, 13–17 March.

Tan, S.C. (2002) 'Water as a bargaining ploy', *New Straits Times*, 15 August.

Tan, T. (2007) 'Newater to meet 30% of needs by 2011', *The Straits Times*, 16 March.

Tan, T.H. (2002a) 'Water: S'pore to rely less on KL', *The Straits Times*, 6 April.

Tan, T.H. (2002b) 'KL-S'pore talks hit snag as Malaysia changes tack', *The Straits Times*, 4 September.

Tan, T.H. (2010) 'Singapore's print media policy: a national success?', in T. Chong (ed.) *Management of Success: Singapore Revisited*, Institute of Southeast Asian Studies, Singapore, pp. 242–256.

Tan, W.T. (1986) *The HDB Resettlement Department and its Role in the Cleaning up Programme of Singapore River and the Kallang Basin Catchments*, COBSEA Workshop on Cleaning-up of Urban Rivers, Ministry of the Environment Singapore and UNEP, 14–16 January, Singapore.

Tan, Y.S., Lee T.J. and Tan, K. (2009) *Clean, Green and Blue: Singapore's Journey towards Environmental and Water Sustainability*, Institute of Southeast Asian Studies, Singapore.

Tang, W.F. (2002) 'Reclaimed water for wafer fabs by next year', *The Business Times*, 23 May.

Tay, K.P. (2008) *Twin Pillars of State Rejuvenation*, Ethos, World Cities Summit Issue, Civil Services College, Singapore, pp. 32–42.

Tebrau and Scudai Rivers Agreement between the Government of the State of Johore and the City Council of the State of Singapore signed on 1 September 1961. Singapore Parliamentary Debates, Official Report, vol. 75, cols. 2605–2644, Annex B, 25 January 2003.

Tee, E. (2000) 'Stanford joins NTU on water project', *The Straits Times*, 24 June.

Teh, H.L. (2002) 'Johor chose not to review water rates: S'pore', *The Business Times*, 28 January.

Teo, A. (1998) 'KL has sent S'pore draft loan agreement: PM Goh', *The Business Times*, 24 November.

Teo, B.T. (1986) *Role of Environmental Health Department in the Clean Up Programme*, COBSEA Workshop on Cleaning-up of Urban Rivers, Ministry of the Environment of Singapore and UNEP, 14–16 January, Singapore.

Teo, G. (2002) 'NEWater can replace Johor supply; DPM Lee says water bought elsewhere must be competitive with reclaimed water, which is a "serious alternative"', *The Straits Times*, 13 July.

Teo, S.E. (1992) 'Planning principles in pre- and post-independent Singapore', *The Town Planning Review*, vol. 63, no. 2, pp. 163–185.

Toh, A. (2002) 'M'sia mulls new laws to dilute water pacts; minister says law could allow Johor to determine amount to supply to S'pore', *The Business Times*, 25 October.

Toh, C.C. (1959) *Legislative Assembly Debates, State of Singapore, Official Report*, First Section of the First Legislative Assembly, Singapore, 16 July.

Toh, E. (1998) 'Suspend fresh ties with S'pore, urges youth chief', *The Business Times*, 4 August.

Toh, E. (2000) 'Johor to build new RM 700m water treatment plant', *The Business Times*, 19 August.

Toh, E. (2001a) 'Johor to have better water management', *The Business Times*, 24 April.

Toh, E. (2001b) 'SM Lee, Dr M strike in-principle accord on bilateral issues', *The Business Times*, 5 September.

Toh, E. (2001c) 'Singapore gave in to Malaysia on bilateral pact: SM Lee', *The Business Times*, 6 September.

Toh, E. (2002a) 'Dr M sent new water proposal to SM last week: DPM Lee', *The Business Times*, 12 March.

Toh, E. (2002b) 'S'pore, KL officials don't see quick breakthrough on water', *The Business Times*, 1 July.

Toh, E. (2002c) 'Malaysia to review retroactively price of water', *The Business Times*, 3 July.

Toh, E. (2002d) 'M'sia seeks up to RM3 per thousand gallons of water', *The Business Times*, 8 July.

Toh, E. (2002e) 'Water price review must be part of a package: PM', *The Business Times*, 9 October.

Toh, E. (2002f) 'Current and future pacts included', *The Business Times*, 17 October.

Toh, E. (2002g) 'M'sia mulls new laws to dilute water pacts', *The Business Times*, 25 October.

Toh, E. (2002h) 'Will M'sia do a "Clob" on water supply?', *The Business Times*, 28 October.

Toh, E. (2003a) 'KL launches ad blitz over water dispute', *The Business Times*, 14 July.

Toh, E. (2003b) 'Another KL-S'pore, cross-border deal', *The Business Times*, 26 July.

Toh, E. (2003c) 'Malaysian govt move could herald new water laws', *The Business Times*, 1 August.

Toh, E. (2003d) 'KL will supply water despite price dispute', *The Business Times*, 2 August.

Toh, E. (2004) 'S'pore firm to dip into China water business', *The Business Times*, 16 June.

Tortajada, C. (2006) 'Water management in Singapore', *International Journal of Water Resources Development*, vol. 22, no. 2, pp. 227–240.

Trade Effluent Regulations (1976) Publication No. SLS.27/76, Government Printer, Singapore.

Trade Effluent (Amendment) Regulations (1977) Publication No. SLS.39/77, Singapore National Printers, Singapore.

Tse, Y.S. (2003) 'Give engineers due credit for contributions', *The Business Times*, 17 March.

Turnbull, C.M. (1977) *A History of Singapore 1819–1975*, Oxford University Press, Singapore.

Turnbull, C.M. (1989) *A History of Singapore (1819–1988)*, Oxford University Press, Singapore.

Turnbull, C.M. (2009) *A History of Modern Singapore: 1819–2005*, NUS Press, Singapore.

United Nations Economic Commission for Asia and the Far East, Division of Water Resources Department (ECAFE) (1964) *Reports on the Industrial Water Supply Project for the Jurong Industrial Estate in the Island*, ECAFE, Bangkok.

United Nations Educational, Scientific and Cultural Organisation (UNESCO) (2006) *2nd UN World Water Development Report*, UNESCO, Paris.

University of California (UCDAVIS) (1996) *Migration News*. Available at: http://migration.ucdavis.edu/mn/more.php?id=929_0_3_0 (accessed on 4 September 2012).

Urban Redevelopment Authority (URA) (1989) *Master Plan for the Urban Waterfronts at Marina Bay and Kallang Basin, Draft*, Urban Redevelopment Authority, Singapore.

Urban Redevelopment Authority (URA) (1991) *Living the Next Lap: Towards a Tropical City of Excellence*, Urban Redevelopment Authority, Singapore.

Urban Redevelopment Authority (URA) (1992) *Singapore River Development Guide Plan, Draft*, Urban Redevelopment Authority, Singapore.

Urban Redevelopment Authority (URA) (1993) *Aesthetic Treatment of 'Waterbodies' in Singapore*, Urban Redevelopment Authority, Singapore.

Urban Redevelopment Authority (URA) (1997) *New Urbanity: The Kallang Basin Redevelopment in Singapore*, Urban Redevelopment Authority, Singapore.

Urban Redevelopment Authority (URA) (1999) *Living the Next Lap: Towards a Tropical City of Excellence*, Urban Redevelopment Authority, Singapore.

Utusan Malaysia (5 July 2002) 'Water issues after payment by Singapore of CPF' (translated from Malay into English).

Utusan Malaysia (9 July 2002) 'Solving water issues opens the door to finalise other outstanding issues' (translated from Malay into English).

Utusan Malaysia (7 August 2002) 'Not keen to extend the 2061 water agreements' (translated from Malay into English).

Utusan Malaysia (8 August 2002a) 'Singapore decision was expected' (translated from Malay into English).

Utusan Malaysia (8 August 2002b) 'People react to cynical move of Singapore towards wastewater recycling' (translated from Malay into English).

Utusan Malaysia (9 October 2002) 'Water: Singapore agrees revisions' (translated from Malay into English).

Utusan Malaysia (19 October 2002) 'Singapore claims that Malaysia did not give complete information' (translated from Malay into English).

Utusan Malaysia (26 October 2002a) 'Cancel the water agreement: Malaysia should be ready to defend the decision' (translated from Malay into English).

Utusan Malaysia (26 October 2002b) 'Water supply issues to Singapore should be studied from all angles' (translated from Malay into English).

Utusan Malaysia (2 November 2002) 'PM: we are entitled to review the price of water' (translated from Malay into English).

Utusan Malaysia (4 November 2002) 'Singapore should be grateful' (translated from Malay into English).

Utusan Malaysia (13 November 2002l) 'Singapore makes RM1.5b from selling water to Malaysia' (translated from Malay into English).

Utusan Malaysia (14 November 2002) 'Raise taxes on the land to Singapore' (translated from Malay into English).

Utusan Malaysia (15 November 2002) 'Water issue: Hamid hopes to negotiate with Jaya-kumar' (translated from Malay into English).

Utusan Malaysia (18 November 2002) 'Singapore-Malaysia water talks resume early next year' (translated from Malay into English).

Utusan Malaysia (28 January 2003) 'Singapore has not matured' (translated from Malay into English).

Utusan Malaysia (29 January 2003) 'Singapore publicised the documents because they need to find a solution' (translated from Malay into English).

Utusan Malaysia (30 January 2003) 'Singaporeans concerned about economic issues' (translated from Malay into English).

Utusan Malaysia (10 February 2003) 'Water law will be enacted' (translated from Malay into English).

Utusan Malaysia (20 February 2003) 'SG must change its kiasu attitude' (translated from Malay into English).

Utusan Malaysia (14 July 2003) 'The people should understand water issue' (translated from Malay into English).

Utusan Malaysia (21 July 2003) 'Do not damage the relationship' (translated from Malay into English).

Utusan Malaysia (1 September 2003) 'Malaysia–Singapore water issues have progressed in secret' (translated from Malay into English).

Utusan Malaysia (3 September 2003) 'Johor will have its own provision of clean water'.

Van Ast, J.A. and Boot, S.P. (2003) 'Participation in European water policy', *Physics and Chemistry on the Earth*, vol. 28, no. 12–13, pp. 555–562.

Vaz, S.J. (1977) 'Why PUB doesn't allow fishing in reservoirs', *The Straits Times*, 18 July.

Venudran, C. (1999) 'Ghani: our need for water to be given priority', *New Straits Times*, 7 June.

Waller, E. (2001) *Landscape Planning in Singapore*, Singapore University Press, Singapore.

Waste News (29 October 2007) 'Speaker highlights new water reuse program'.

Water Pollution Control and Drainage (1975) Act 29 of 1975, Singapore National Print-ers (Pte) Limited, Singapore.

Water Pollution Control and Drainage Act (Chapter 348) 1975, Government Printer, Singapore.

White, B. (1950) *Report on the Water Resources of Singapore Island, Excluding Those within the Present Protected Catchment Area*, Wolfe Barry & Partners, London.

White, B. (1952) *Government of Singapore: The Water Resources of Singapore Island; Report on the Development of the City of Singapore Water Supply and Emergency Supplies in Relation Thereto*, White (Sir Bruce), Wolfe Barry & Partners, London.

Wong, C. (1998a) 'Water pact with Singapore just does not add up', *New Straits Times*, 21 July.

Wong, P.P. (1969) 'The changing landscapes of Singapore island', in J.B. Ooi and H.D. Chiang (eds) *Modern Singapore*, Singapore University Press, Singapore, pp. 20–51.

Wong, R. (1998b) 'M'sia seeks Singapore's help to raise funds', *The Business Times*, 6 November.

Wong, T.C., Yuen, B. and Goldblum, C. (eds) (2008) *Spatial Planning for a Sustainable Singapore*, Springer, Singapore.

World Bank (1992) *Water Supply and Sanitation Projects: The Bank's Experience, 1967–1989*, Report no. 10789, World Bank, Washington, DC.

World Bank (2010) *World Development Indicators Database 2008*. Available at: http://siteresources.worldbank.org/DATASTATISTICS/Resources/GNIPC.pdf (accessed on 20 July 2010).

World Health Organisation (WHO) (2006) *Guidelines for Drinking Water Quality*, (incorporated as the First Addendum), 3rd Edition, vol. 1, World Health Organisation, Switzerland.

Xinhua (12 January 2001) 'No threat of water cut-off from Malaysia: S'pore minister'.

Xinhua (12 July 2002) 'Singapore won't agree to one-sided water agreement with Malaysia: deputy PM'.

Xinhua (25 September 2001b) 'Singapore-Malaysia agreement: a psychological breakthrough: FM'.

Xinhua (1 August 2003) 'Malaysia's new move not to affect water supply to Singapore: PM'.

Xinhua (23 November 2004) 'Singapore to become global hydro hub'.

Yang, R.K. (1998) 'Anwar clarifies changes on Bumiputra ownership rules', *The Business Times*, 10 April.

Yap, C.G.E. (1988) *Budget, Ministry of Trade and Industry*, Parliament no. 6, Session no. 2, vol. 50, Sitting no. 18, Sitting date: 25 March, Hansard, Singapore.

Yap, K.G. (1986) *Physical Improvement Works to the Singapore River and the Kallang Basin*, COBSEA Workshop on Cleaning-up of Urban River, Ministry of the Environment of Singapore and UNEP, 14–16 January, Singapore.

Yap, S., Lim, R. and Kam, L.W. (2010) *Men in White: The Untold Story of Singapore's Ruling Political Party*, Singapore Press Holdings, Singapore.

Yeo, C.T. (1995) *Debate on Annual Budget Statement*, Parliament no. 8, Session no. 2, vol. 64, Sitting no. 2, Sitting date: 13 March, Hansard, Singapore.

Yeoh, E. and Ting, L.H. (2000) 'Massive water project is floated', *The Straits Times*, 2 July.

Yeung, Y. (1973) *National Development Policy and Urban Transformation in Singapore: A Study of Public Housing and the Marketing System*, University of Chicago Press, Chicago.

Yew, F.S., Goh, K.T. and Lim, Y.S. (1993) 'Epidemiology of typhoid fever in Singapore', *Epidemiology and Infection*, vol. 110, no. 1, pp. 63–70.

Yue, A.Y. (1973) *National Development Policy and Urban Transformation in Singapore: A Study of Public Housing and the Marketing System*, University of Chicago Press, Chicago.

Index